T0150816

NUMB
AND
NUMBER

NUMB AND NUMBER

HOW TO AVOID BEING MYSTIFIED BY
THE MATHEMATICS OF MODERN LIFE

WILLIAM HARTSTON

Atlantic Books
London

First published in Great Britain in 2020 by Atlantic Books,
an imprint of Atlantic Books Ltd.

1 2 3 4 5 6 7 8 9

A CIP catalogue record for this book is available from the British
Library.

Hardback ISBN: 978-1-83895-084-2
E-book ISBN: 978-1-83895-086-6
Paperback ISBN: 978-1-83895-085-9

Printed in Great Britain

Atlantic Books
An Imprint of Atlantic Books Ltd
Ormond House
26–27 Boswell Street
London
WC1N 3JZ

www.atlantic-books.co.uk

**Do not worry about your difficulties in Mathematics.
I can assure you mine are still greater.**

(Albert Einstein, 1943, in a letter replying to a little girl
who was having problems with maths at school)

Contents

Introduction   1

1. The Number of Our Days
   How life expectancy confounds our expectations   13

2. Surveying the Scene
   How to mislead with opinion polls   33

3. Risk and Behaviour
   Our illogical attitudes towards risk   53

4. The Mathematics of Sport
   The randomness of winning and losing   71

5. Saved You!
   How governments try to look good   102

6. Numbers Large and Small
   How huge and tiny numbers confuse us   122

7. The Insignificance of Significance
   The misleading language of statisticians   133

8. Cause and Effect
   Common logical confusions   154

9. Percentages and More Misleading Mathematics
   More natural mistakes   175

10. Chaotic Butterflies
    The mathematics of chaos, catastrophe and
    complexity   193

11. Torpedoes, Toilets and True Love
   A difficult problem with many applications   217

12. Formula Milking
   Unbelievable formulae in newspapers   232

13. Monkey Maths
   An evolutionary perspective on numeracy   241

14. Pandemic Pandemonium
   The world's reaction to coronavirus   248

Bibliography   271

Acknowledgements   275

Index   276

Introduction

'It has become almost a cliché to remark that nobody boasts of ignorance of literature, but it is socially acceptable to boast ignorance of science and proudly claim incompetence in mathematics.'

(Richard Dawkins, BBC Richard Dimbleby Lecture, 1996)

One morning in December 2019, amid the deluge of opinion polls and electoral statistics in the British newspapers, TV current affairs programmes and my email inbox, I was particularly intrigued by two news reports and one press release.

The press release informed me that 10 million people in the UK suffer from headaches regularly, while the news stories carried the information that 37% of children in Croydon live in relative poverty and that exports from China to the United States in November fell 23% from a year earlier.

We live in a world in which the news has become increasingly dominated by data and numbers, but all too often they are flung at us with insufficient information to interpret them properly. The figures may cause some concern for people with headaches, for child welfare services in Croydon and for Chinese exporters, but most of us, including those groups, rarely stop to think about what they really mean and how they were calculated. We numbly let the numbers wash over us and do not pause to consider such questions as the following:

What do they mean by 'suffer from headaches regularly'? Do they mean 'frequently'? Once a year is regular but it is not frequent, as headaches go. This reminds me of a sign I have occasionally seen at railway stations advertising a 'regular service to London'. One train a week may be regular, but that's hardly a good advert for the service. And if those headaches are suffered frequently, does this mean every day, or every morning after a heavy drinking session, or whenever work piles up, or what?

And what about those Croydon children living in relative poverty? What is 'relative' meant to mean? Surely the 37% of children who are poorest are relatively poorer than the 63% who are richer than them? And are they poor relative to other children in Croydon or are we comparing them with the rest of the country? Without defining what is meant by 'relative poverty', giving a precise figure such as 37% is meaningless.

Finally, what should we make of that drop in exports from China to the US? When they say 'fell 23% from a year earlier' does that mean comparing the monthly figures for November 2019 and 2018, or comparing the figures for the whole year ending in those months? Also, there have been considerable fluctuations in exchange rate between the US dollar and Chinese yen over the year. Do we get the same 23% result if we compare the figures in Chinese currency? And what about exports from the US to China?

While we are on the subject, we should always pause to consider what it really means when one year's figures, whether they are measures of trade, profits, turnover or anything else, are compared with those of the previous year. Were that year's figures particularly good or very disappointing? Was this year's 23% fall an expected counterbalance to last year's rise, or did it just continue, or even accentuate, a path that had been declining for some years?

Most of us have neither the time nor the inclination to seek the answers to such questions but we should at least be aware that the figures flung at us may not tell the whole story. Particularly at election times, politicians, advertisers and others who seek to influence us carefully select the numbers that support their arguments.

As Lewis Carroll put it, 'Long and painful experience has taught me one great principle in managing business for other people: viz., If you want to inspire confidence, *give plenty of statistics*. It does not matter that they should be accurate, or even intelligible, as long as there is enough of them' (Lewis Carroll, *Three Years in a Curatorship by One*

Whom It Has Tried, 1886). When people detect a deluge of numbers about to fall on them, they all too often put up their numbrellas to shelter from the figures and formulae and leave them falling unheeded to the ground. This book is an attempt to remove the numbness that numbers often induce while retaining a proper sense of scepticism when figures and statistics are flung in our direction.

Every science goes through three stages in its evolution: superstition, empiricism and finally mathematics. Chemistry had its roots in the theory, first proposed by Empedocles in the fifth century BC, that everything is composed of the four elements: earth, air, fire and water. Medieval alchemists introduced experimentation, and even Isaac Newton called himself an alchemist when looking for the mathematical rules behind physics and chemistry. When Aristotle, in the fourth century BC, explained why children look like their parents, he put it down to the mother's imagination: 'If in the act of copulation, the woman earnestly look on the man, and fix her mind on him, the child will resemble its father. Nay, if a woman, even in unlawful copulation, fix her mind upon her husband, the child will resemble him though he did not beget it.'

This mistaken explanation may have laid the foundations for genetics, which only entered the age of empiricism with the experimental breeding of peas by the Augustinian friar Gregor Mendel in the 1860s. It took almost another century before the discovery of DNA turned genetics into a proper mathematical discipline.

Until the year 1834, the word 'scientist' was unknown in English, and the first appearance of 'genetics' was in 1872. Yet, despite the developments of modern science and the public's increasing dependence on it, far too many of us are content to remain in the realm of superstition, with occasional forays into the empiricist's experimental methods, without even being tempted to venture into the realm of mathematics which can explain it all. We'd like to understand but are put off by the abstraction of mathematics, which makes the whole subject a no-go zone.

I suspect that the real trouble is that we process language so fast that, when numbers come along, we need to slow down in order to take them in and we are reluctant to do so. Mathematics in general and numbers in particular have become mere adornments, included in news stories and policy documents to add apparent weight to whatever is being said and convey a feeling of scientific validity.

When we hear reports of primitive societies that have no words to express numbers greater than three, we smile smugly and think that 'one, two, three, many' is no way to count, yet we have our own problems as far as large numbers are concerned. We are mostly all right with four and five, and have a fair grasp of hundreds and thousands, but once we get up to millions, or billions, or trillions, we become bemused. What's the difference between a billion and a trillion anyway? We see both of them as 'many' so shrug it off when we hear that the UK national debt is over £1.8 trillion. Only when we learn that £1.8 trillion is more than £27,000 for every person in the country do we begin to understand just how much it is.

In this book I try to explain some of the ideas behind the figures and formulae we encounter every day and some of the maths behind them. We shall reveal many dubious ways in which statistics may be used by politicians and advertisers to mislead us, and the manner in which writers of film scripts and the press may mangle beautiful mathematical ideas such as Chaos Theory to inflict on us a horribly simplified version that gives the wrong idea entirely.

My underlying message is that mathematics is beautiful and provides the necessary tools to help understand the world around us and adopt a rational approach to the problems of life. I hope that this book will help readers to overcome any fear of figures they may harbour, and the next time they are caught in a deluge of numbers, statistics and mathematical symbols they will keep their numbrellas tightly furled and welcome all the numbers with an open and better-informed mind.

For me, I suppose, many of the ideas in this book really began with walruses. Those marine mammals wobbled their way into the picture long after my ill-spent youth studying mathematics and playing too much chess. After spending my formative years in that way, I started becoming interested in the rest of the world and began reading newspapers. That was when I discovered that the vast majority of people have a different reaction to numbers from that of mathematicians. Whenever I came across a statistic or other numerical data in a news story, I paused and asked what it really meant and that question more often than not left me with a feeling of dissatisfaction or even bemusement. The verbiage around the number usually provided insufficient information about the data they were relying on for me to assess or even understand the results.

I came to realize that most people, including the journalists responsible for the newspaper stories and almost all the people reading them, just saw the numbers as background adornments, letting them float past without consideration, trusting that they told the same story as the words. The numbers induced a comforting numbness that reassured the reader. But when I read about the walruses, I was far from reassured.

The story, which appeared in several UK newspapers, concerned an intriguing piece of research published in the journal *BMC Ecology* in 2003 under the title 'Feeding Behaviour of Free-Ranging Walruses with Notes on Apparent Dextrality of Flipper Use'. The sentence in the paper that all the reports latched on to was one reporting that during feeding behaviour, 'the walruses used their right flipper 89% of the time' but interpretations of that sentence differed.

Some concluded that it meant that 89% of walruses were right-flippered and commented on the fact that this was much the same as the percentage of humans that are right-handed. Others reported that the research had found that all the walruses in the study were right-flippered and used that flipper 89% of the time compared with the left flipper 11% of the time.

Quite apart from wanting to know which of those was the case, I also became intrigued by that 89% figure for right-handedness in both walruses and humans and I asked myself whether that was really what the figures showed. If 89% of walruses are right-flippered, would we expect 89% of recorded flipper uses to involve the right flipper?

I did a quick calculation and found that this was surprisingly not the case. Let's make some reasonable assumptions to keep the sums easy and see what happens.

Various pieces of research have come up with slightly different figures for the proportion of humans who are right-handed and the results vary between about 85% and 92%. So, to keep the calculation simple, let us suppose that 90% of us are right-handed and all of us use our dominant hand 90% of the time.

Then, in a sample of 100 people, we have on average 90 right-handers and 10 left-handers. Giving each of them ten trials, our right-handers will each use their right hand nine times and left hand once, while the left-handers will do the reverse. Our 90 right-handers will therefore give a total of 810 uses of the right hand and 90 uses of the left hand, while the 10 left-handers will have 90 uses of the left hand and 10 of the right hand. Our observations will therefore comprise a total of 820 right-hand uses and 180 left-hand uses. Paradoxically, it may seem, our assumption of a 90%–10% split in handedness has led to an 82%–18% split in observations. So walruses and humans, as far as handedness is concerned, are not the same after all.

I read the paper on which all these walrus reports were based and I contacted one of the authors to ask what the results had really shown. It turned out that the research was based on video footage of walruses feeding in Greenland, each video cut into fixed-length segments and each segment examined for flipper use. Only segments showing preferential use of one flipper over the other were included in the analysis, which formed the basis for the 89% figure. The number of walruses involved in the study was unknown, though the researchers told me that it was at least five.

Five walruses, however, do not constitute a significant sample (as we shall see when we discuss sample size and significance later in this book) and, as the researchers said, further research was clearly needed.

Sadly, I have not come across any later research into handedness in walruses, other than a 2014 report of the examination of the tusks of seventeenth-century walrus skeletons in Nova Scotia which concluded that walruses may have been predominantly left-handed. The mystery of handedness in walruses thus remains unresolved but at least we now know a little better what questions we should be asking of any observational or experimental results.

Finally, however, and to add to the confusion, I should add something about another piece of research into the handedness of a different sea creature that I learnt about just as I was writing about walruses. At the end of November 2019, the *Guardian* newspaper ran a story about newly published research under the headline: 'Most dolphins are "right-handed", say researchers'. A BBC World Service news report used exactly the same words and this was eagerly reported in other media too, but quite apart from the fact that dolphins do not have hands, the conclusion is not as evident as they made it sound.

The research on which it is based was published in November 2019 under the title 'Behavioural Laterality in Foraging Bottlenose Dolphins (*Tursiops truncatus*)' in the *Royal Society Open Science* journal and detailed the results of observing foraging behaviour by dolphins off Bimini in the Bahamas. This behaviour, the authors explain, involves dolphins swimming slowly along the ocean floor using echo location to identify a potential food source. When they have found one, they turn sharply, bury their rostrum (beak) in the seabed and dig out the food.

Analysing 709 such turns by at least 27 distinct dolphins, the authors report that 705 were to the left, with their right side and right eye downwards. Furthermore, all four examples of right turns were produced by the same dolphin which, the researchers point out, had 'an abnormally shaped right pectoral fin' which may have affected its turning behaviour.

The news reports of the research did not go into the reasons for concluding that left turns indicated right-handedness, but the researchers point out that previous research had indicated that a dolphin's sight is better through its right eye and the echo-location clicks it makes are produced better from their right-hand 'phonic lips' than those on the left.

It may therefore make sense for a dolphin to keep its right side facing downwards where the food is located, but it seems to me it is just as easy, if you happen to be a dolphin, to keep your right side downwards when turning right as it is when turning left.

As the researchers point out, many animals have been shown to prefer one side to the other. Chimpanzees and gorillas show a significant right-hand bias, but orangutans are mainly left-handed. Herds of reindeer tend to circle in an anticlockwise direction while giraffes when splaying their legs tend to move their left leg first. For many more species, however, including lions, bats, chickens, parrots and toads, individual animals show a preference for using their right or left limbs, but there is no significant overall species preference for either right or left.

A good deal of research shows that newborn babies have a greater tendency to turn their heads to the right and this is seen by some as due to our greater right-handedness (though it may also be a sign of preferential muscle development). It is not clear to me why turning to the left is seen as a sign of left-handedness in babies but right-handedness in dolphins.

Despite reputable research showing that polar bears use their right and left paws equally, many collections of trivia or supposedly surprising information include the 'fact' that 'all polar bears are left-handed'. This particular piece of disinformation can apparently be traced back to a single account of one polar bear seen by one Native American chieftain covering its nose with its right paw when sneaking up on a sea lion before battering it with its (preferred) left paw. Five walruses was a small sample in the earlier study but a sole polar bear is an even more extreme example.

6	∞	9	3	8
2	7	=	1	%
4	¾	3	2	±
Σ	9	7	¼	3
¾	0	x	4	8

CHAPTER 1

The Number of Our Days

How life expectancy confounds our expectations

Show me, Lord, my life's end and the number of my days.

(Psalms 39:4, The Bible, New International Version)

Trying to estimate the number of our days is an essential part of life insurance and pension planning, but the calculation of life expectancy is something most people do not think about, and even among those who do think about it, very few understand.

Life is invariably fatal: 100% of people die. Or do they? It is frequently claimed, by people who do not pause to think about what they are saying, that more people are alive today than have ever died. That's nonsense. There are currently, according to UN estimates, just over 7.7 billion people on Earth. Lack of accurate data, or indeed any data at all for much of the time, makes it difficult even to guess the number of people in past times, but the table gives plausible estimates for the years in which landmark figures were reached.

Year	World Population
2011	7 billion
1999	6 billion
1987	5 billion
1974	4 billion
1960	3 billion
1927	2 billion
1804	1 billion
1700	610 million
1600	500 million
1500	450 million
1400	350 million
1100	320 million
800	220 million
600	200 million

From these figures, we may estimate that in the 500 years between 600 and 1100, well over a billion people (5 × 200 million) died, as very few people lived more than a hundred years. Another billion deaths would have been exceeded between 1100 and 1400, followed by more than 2 billion from 1400 to 1800. When we add the billion alive in 1804 and the 2 billion in 1927, we are already well over 7 billion.

Homo sapiens emerged between 50,000 and 300,000 years so, which adds a large number of dead people to our collection. We can only guess at population sizes and average lifespans in the early years but the best-informed estimates have reached the conclusion that around 108 billion people have ever lived. This means that about one-fourteenth of the people who have ever lived are alive today.

Looking on the bright side, this means that only 13 out of every 14 people who have ever been born have died so we might optimistically conclude that we have a 1 in 14 chance of living forever. Right?

Er, no. Just wait another 100 years or so and we'll have the complete data on almost all of us. And that introduces the real problem of estimating life expectancy.

Having complete data is a problem often ignored by over-eager users of medical statistics. What should we make, for example, of a recent report that deaths from breast cancer have been going down by almost 2% per year? Since everyone dies, a reduction in deaths from one cause must be matched by an increase in others. A decrease in premature deaths means something; an overall decrease needs further probing. And what does it mean when we read that life expectancy at birth in the UK is 79.6 years? It certainly doesn't mean that the average age at which British people die is 79.6, because anyone who is 79 years old was

Extrapolation

In bygone years, if we wanted to predict the future we would look at stars in the sky, or deal out Tarot cards, or look for patterns in tea leaves or animal entrails. Now we collect vast amounts of data, plot them on graphs, try to detect patterns and work out what will happen if those patterns continue into the future. That's extrapolation. On the whole, it gives more reliable results than Tarot cards or entrails, but as we see with life expectancy figures, trying to predict long into the future is a far from exact procedure. One cannot, after all, extrapolate progress.

Sometimes statisticians, especially those of an ultra-cautious disposition, refer to their suggestions of future trends not as predictions but projections. Predictions based on statistics are always made, to some extent at least, on the assumption that the future will work in much the same way as the present. Using the word 'projection' stresses that point and, in the case of life expectancy calculations, it is an admission that any attempt to extrapolate medical progress a lifetime ahead is little more than educated guesswork.

born 79 years ago, not today. To predict how long a person born today will live involves making a prediction of medical progress for the next century. The spurious accuracy of that 79.6 figure covers a calculation based on a variety of assumptions, many of which are not easy to justify. Yet published figures of life expectancy have a huge economic effect on pension funds and government planning. Before considering the present state of life expectancy calculations, however, let us go back to its beginnings.

The earliest known collection and publication of data in a form that resembled a life expectancy table was by the Roman jurist Ulpian around AD 220. Much admired at the time as a legal authority, he advised on a system of taxation and inheritance payments that involved a death tax of around 5% on any legacy, with the remaining 95% funding an annual payment to the recipient of the legacy at a prescribed rate in a manner similar to modern annuities. To determine a fair rate, however, an estimate was needed of the life expectancy of the recipient and that was what Ulpian's figures set out to provide.

Where his figures came from is not known, and great doubts have been expressed concerning their reliability and the statistical methods used for their calculation, but they suggest a female life expectancy at birth of 22.5 years and a male life expectancy of 20.4. For anyone reaching their late 30s, however, Ulpian predicts another 20 years of life while the over-60s could count on another five years on average. Ulpian himself lived to his early 50s. He was murdered in AD 223 in a riot between the soldiers and the mob by members of the Praetorian Guard, whom he had annoyed by reducing their privileges some years earlier.

Ulpian's tables remained the last word in life expectancy predictions in the Roman Empire for several hundred years and were not surpassed significantly until the seventeenth century when an Englishman, whose name is now associated with a less earthly, more celestial observation, had a very bright idea. That man was to become England's second Astronomer Royal and his name was Edmond (sometimes spelt Edmund) Halley.

Quite apart from his discovery of the comet named after him and his correct prediction of its return in a 76-year cycle, Halley made prodigious contributions to a number of scientific fields from an early age. He went to Queen's College, Oxford, at the age of 16 and published papers on sunspots and the Solar System while still an undergraduate. He left Oxford after four years, without having taken a degree, to set up an observatory on the island of Saint Helena. Oxford made up for his formal failure to graduate by giving him an MA degree when he was 22, at which age he was also elected to be a Fellow of the Royal Society. He died at the age of 86, in 1742, allegedly after drinking a glass of wine against his doctor's orders.

Halley's great contribution to the science of life expectancy came in 1693 when he had been working in the Austrian town of Breslau (now Wrocław in Poland). He had come across data recording the annual numbers of births and deaths in the town over a five-year period and, most importantly, the sex and age of those who had died. Since Breslau was a small, tightly knit community far from the sea, and the number of births was roughly equal to the number of deaths over the five-year period, Halley reasoned that the number of people joining or leaving the town each year, other than through births and deaths, was small. This meant the total population (for which the records gave no precise figure) was reasonably stable, and that assumption enabled Halley to draw far-reaching conclusions.

His method was simple: the assumption of a constant birth rate told him the number of people of any age who would be alive if none of them had died, and by summing the mortality numbers for all lower ages, he calculated how many were still alive. This technique enabled him to calculate the odds an individual of any specified age had of reaching his next birthday. His table, for example, gave the number of 25-year-olds living in Breslau as 567 while the number of 26-year-olds was 560, so in his own words: 'As for Instance, a Person of 25 Years of Age has the odds of 560 to 7 or 80 to 1 that he does not die in a year: Because that of 567, living 25 years of Age there die no more than 7 in a year, leaving 560 of 26 Years old.'

Halley did not specifically calculate life expectancy at birth, but we can see from his figures that around 50% of people died before they were 34. He did, however, devote a good deal of space to the calculation of sensible rates of annuities, as he made clear in the full title of his report: 'An estimate of the degrees of the mortality of mankind; drawn from curious tables of the births and funerals at the city of Breslaw; with an attempt to ascertain the price of annuities upon lives.'

In fact, much of the motivation behind Halley's work in this respect lay in the English government's policy of raising funds for the war against France by selling annuities but the rates they offered were considerably less justifiable than those of Ulpian back in ancient Rome.

At one stage, the English government offered annuities that paid back their full price in only seven years; even when that time frame was doubled to 14 years, no account was made for the age of the purchaser. Nowadays, annuity rates are based on mortality tables. The total amount invested is usually built up over an individual's working life, then used to purchase the annuity, which guarantees to pay a certain amount each year. The word 'annuity' comes from *annus*, the Latin for 'year', and the annual amount depends on the age of the individual and the number of further years they can expect to live. Back in the seventeenth century, the economist William Petty and the statistician John Graunt had made valiant efforts to draw up mortality tables for London some 30 years before Halley came along, but they lacked the precise data to calculate life expectancy that the town of Breslau offered, so their figures offered only limited help in making financial decisions. Without their contributions, however, Halley would probably not have had the idea of using the data in the way he did.

Today, it is clear that statisticians still rely on the basic techniques introduced by Halley, though the pace of change in modern life requires some significant changes and additions to those techniques.

When the Office for National Statistics tells us that life expectancy at birth in the UK is 87.6 years for a man and 90.2 years for a woman, what exactly does that mean and how did they calculate it?

Averages: Mean, Median or Mode?

'Average' is potentially a misleading word, for there are three common mathematical types of average: mean, median and mode. Which one is being used is rarely specified, but it can make a big difference.

When most people mention the average of a group of numbers, they are referring to the mean, which is what you get when you add all the numbers together then divide by the number of values in the group.

For the median, we put all the values in order from smallest to largest and the median is the one in the middle.

Finally, the mode (or modal value) is the most common value in the group.

The mean age at which men died in the UK in 2019 was 79. The median age of those deaths was 82 (so, of those who died, the same number of people were under 82 as over 82). The mode (the most common age) was 85. (For all three averages, the figures for women were around three years higher.)

Until around 1840, the modal age for people dying in the UK was 0, because of very high mortality rates shortly after birth. Remarkably, even in most developed countries, until the second half of the twentieth century it remained the case that more people died before their first birthday than during any other specific year of their life.

Life expectancy, you may think, is exactly what it says: the number of years we can expect to live. Well no: it's not really that at all. Nor is it the mean average age at which people die, the median age at which people die (meaning that about half of all deaths occur before that age and half after) or the most common (modal) age at which people die. We shall come to what life expectancy really means in a moment, but first let's look at a few figures.

In the three-year period 2016–18 in the UK, the mean age at which people died, which is what most people mean when they refer to an 'average', was 79.3 for men and 82.9 for women but the median age at death was 82.5 for men and 85.9 for women.

At first sight, this discrepancy between mean and median may seem strange, but it is just what one would expect when you consider the nature of the data. The ages at which people die spread from zero upwards, and the lower figures have a strong influence in pulling down the mean. A person may die 80 years before reaching the mean, but nobody reaching it lives another 80 years. All death ages count the same towards the median figure, so the median ends up higher than the mean.

This is further emphasized by the figures for the modal age of death – the age at which most deaths occur, which one could consider the 'typical' or most likely age of death. In 2018, for men in the UK this was 86 and for women it was 88. The fact that the difference in average age at death between men and women is only 2 years for the mode, while for the median it is 3.4 and for the mean it is 3.6, seems mainly due to higher infant mortality figures in males.

In 1974, UK mortality figures registered modal ages of 81 for women and 74 for men. The large decrease in the difference since then has been put down to improvements in working conditions for men and considerable decrease in smoking, both of which have greatly increased the lifespans of men.

While all these figures have undergone strong and steady improvement in the last half-century, the greatest change for men has been in the modal age of death – the age at which most deaths occur. In 1967, it was 67, rising to 72 the following year and 74 in 1969. It remained in the mid-seventies until 2000 when it increased to 79, then the following year it was 80 and it has slowly crept up to 86 in the most recent figures.

Perhaps most remarkable of all, in the UK the modal age of male deaths until 1966 was zero: infant mortality was responsible for more male deaths before the baby had reached his first birthday than were registered at any later year of life. For reasons that are still unclear, girl babies have a better survival rate than boy babies: the modal age at death for UK females has not been zero since the 1940s.

However, it has never been considered advisable to tell people that their modal life expectancy is zero or that that is the age at which people are most likely to die: starting with Edmond Halley, more useful measures were developed to give a figure that was believable, useful and not so depressing. Whether these measures are generally understood or even intelligible to most people is another matter.

When actuaries draw up tables intended for the use of life assurance companies or long-term governmental planning concerning matters such as pensions and retirement age, such tables usually take the form of lists giving the number of years of life a person may expect at various ages. The Office for National Statistics in the UK (www.ons.gov.uk) has even developed an online Life Expectancy Calculator into which anyone can enter his or her age and sex, click on the 'Calculate' button and it will tell them what age they can expect to live to. A 73-year-old male, for example, will discover that he can expect to live to around the age of 87, which is the same age predicted for a 45-year-old female. A woman aged 97 will discover she has an even chance of living to 100.

Looking at these figures quickly reveals the real problem: they involve seeing into the future. In the case of the 73-year-old man and the 97-year-old woman, we are looking only 14 and three years ahead respectively, so there is a fair chance that our estimates are not too far off. For the 45-year-old woman, however, the prediction looks 42 years into the future and we really have little idea of what global disasters or medical breakthroughs may happen over that period to throw our estimates way off balance.

The UK Office for National Statistics stresses that such figures are 'projections' not 'forecasts', meaning that they project current data into the future, and they have two distinct ways of doing this.

The first is called 'period life expectancy' and is the simpler and probably less accurate of the two. Period life expectancies use current mortality rates as calculated from existing data and assume that those rates apply throughout the remainder of a person's life. This is essentially what Halley did in Breslau. Knowing how many babies were born and how many died in their first year of life enables you to work out the chance that a person will survive from nought to one. Similarly, you work out how many one-year-olds survive to see their second birthday, and how many of those reach the age of three and so on. Eventually you reach an age to which 50% of people have survived and that is taken to be the life expectancy.

Period life expectancies calculated in this manner are precise and based on current data, but possible future changes to mortality rates are not taken into account.

The second method, called cohort life expectancy, is more complex and probably more likely to be realistic, but it involves making certain assumptions about how long into the future historical trends will continue. Such judgements are always liable to have a subjective element.

To assess the cohort life expectancy at birth of a person born in 2020 we would ideally like to know the chance of such a person reaching their first birthday, then the chance of a one-year-old in 2021 reaching the age of two, and a two-year-old in 2022 reaching the age of three and so on. We do not yet, of course, have any of those figures but we have them for previous years. The 'cohort' that gives this system its name is the group of people born in the same year.

Halley's assumption in Breslau was that the probabilities involved in this calculation were stable, so period life expectancy (assuming current trends would continue unchanged) would be the same as cohort expectancy (which tries to take account of changes in death rates). As lifespans have been increasing, lifestyles have changed and infant mortality has been cut immensely as part of a vast increase in life-lengthening medical improvements, we need to massage the figures a little to take such things into account and that is where the clever part of the calculation comes in.

As Edward Morgan of the Office for National Statistics explained in a December 2019 release:

For the 2018-based projections, we set a target mortality improvement rate of 1.2% per year for males and females at most ages in 2043 (the 25th year of the projection) and all future years. Improvement rates for earlier years are obtained by interpolating between the current mortality improvement rate and the target rate. The resulting improvement rates are used to produce projected future mortality rates for each year of the projection. These in turn are used to produce projected life expectancies.

In other words, they looked at rates at which mortality had improved in previous years to produce the figure of 1.2% per year from 2043 onwards, then filled in the 25-year gap between 2018 and 2043 in a smooth way and used the resulting figures to calculate the life expectancies.

The inbuilt improvement rate naturally gives longer cohort life expectancies than would be obtained with period expectancy methods, and just in case anyone thought the assumption of 1.2% annual improvement was wrong, an even higher cohort estimate was produced with annual 1.9% improvement after 25 years, and a lower one with no improvement at all after 25 years. The three variant improvement rates of 0, 1.2% and 1.9% result in life expectancy projections for people born in 2043 of 80.9, 90.4 and 96.3 years for men and 84.1, 92.6 and 98.1 years for women.

In general, for people born around 1950, the gap between

period and cohort life expectancies has been about nine or ten years for both men and women.

The real problem is that however detailed the system of calculation is, saying that a man's life expectancy is 87.6 years at birth is looking almost 90 years into the future which raises the question of whether giving such a precise figure is justified. One decimal point represents one tenth of a year, which is just over one month. United Nations data, however, give figures to two decimal places, which implies accuracy to within three or four days. That is a spuriously impressive claim for figures looking almost a century ahead.

Finally, before we leave life expectancies, let's try to get to grips with some of the claims made about human lifespans long ago and some of the misconceptions that have arisen from them.

There is no doubt that life expectancies around the world have vastly increased, particularly in the last century. It is difficult to say with any degree of certainty how long the average human lived centuries ago as accurate records only began to be kept around the middle of the nineteenth century, but all the available evidence suggests that, even as recently as 1800, the life expectancy in every country was less than 40 years.

On hearing that, we should not fall into the trap of thinking that many people were dropping dead in their 30s or 40s or that very few people reached what is now considered old age, because the main contribution to low life expectancy was infant and child mortality. Even as late as 1950, it is estimated that 16% of babies worldwide died in their first year and 27% did not reach their 15th birthday. Thanks to vast improvements in healthcare, those figures have now dropped to 2.9% and 4.6% respectively.

In England and Wales in 1851, 30% of people did not survive their first ten years, yet if a person survived their childhood, they had a good chance of reaching their 60s. Indeed, although life expectancy at birth was less than 40, even a five-year-old could expect to reach the age of 55.

Throughout Europe between 1200 and 1745, precise figures are scanty, but such evidence as exists suggests that during this period a 21-year-old could expect to live to an age between 62 and 70, except in the fourteenth century when bubonic plague reduced life expectancy for a 21-year-old to 45.

The important thing here is to make a distinction between life expectancy and lifespan. The former is a measure of how long a person will live on average, the second is a measure of how long you will live before you die of old age. The evidence is that human lifespan has increased over the centuries at a far slower rate than life expectancy.

Occasionally claims are made that, in certain societies long ago, lifespans were much the same as they are now. For example, in 1994 a study was made of all people with entries in the *Oxford Classical Dictionary* who had not suffered violent, untimely deaths. That produced a list of 298 ancient Romans all of whom had been born before 100 BC and it was reported that the median age at which they had died was 72. For purposes of comparison, a similar study was performed on people with entries in *Chambers Biographical Dictionary* who had died between 1850 and 1949 and their median age turned out to be 71.

This study led to several reports claiming that the ancient Romans, once they had reached adulthood, lived for much the same time as we do today, but that conclusion falls into another trap: in such cases we should always ask whether the data is representative of the group we are describing and whether the comparison we are making is fair. People with entries in the *Oxford Classical Dictionary* consisted of the wise and the powerful, and in Roman times such status would generally exclude the young. For much of the period under consideration, a man could not even stand for office before he reached the age of 30 and anyone wanting to be appointed a consul had to be at least 43.

By analogy, to take a more extreme example, the average age at death of all popes between 1503 and 1700 was 70 and between 1700 and 2005 was 78 but this does not tell us that human lifespan has been in the 70s for half a millennium. The average age of election for a pope during these periods was 64, so it is hardly surprising that they did not die young. Like Roman consuls, popes are not a representative sample of the population.

The ancient Roman fallacy was brought into perspective in 2016 when the medical historian Valentina Gazzaniga studied some 2,000 skeletons unearthed during construction of a high-speed rail line between Rome and Naples. They had all been buried in mass graves between

Predictions

Everything we have said in this chapter has been concerned with the uses and limitations of statistics in predicting how long we shall live. Astrology, fortune-telling, horserace tipping, investment advice and looking for trends in statistics are all forms of prediction. Some are more scientific than others but behind all of them is a belief that, to some extent at least, the future may be predicted.

One of the troubles with predictions is that they may to a certain extent be self-fulfilling or self-negating. Any supposed 'risk' we claim to identify is a prediction which is liable to alter people's behaviour, thereby reducing the actual 'risk'. Equally, any fortune-teller who predicts that we shall meet a tall, dark, handsome stranger who will change our lives for the better is liable to make us more alert and responsive to the presence of tall, dark, handsome strangers.

the first and third centuries AD, which classified them as labourers from the lower classes of society. That impression was reinforced by the widespread prevalence of arthritis and the high incidence of serious injuries displayed by the bodies. Their average age at death was revealed to be around 30.

The true figure for adult life expectancy in ancient Rome probably lies somewhere between the 30 years of bad diet and hard labour ending in a paupers' grave and the 72 years of patrician success that earned an entry in a Classical Dictionary.

6 ∞ 9 3 8
2 7 = 1 %
4 ¾ 3 2 ±
Σ 9 7 ¼ 3
¾ 0 x 4 8

CHAPTER 2

Surveying the Scene

How to mislead with opinion polls

3 √ 5 1 +
6 4 9
¼ 0 7
5 % 3
0

**At any given moment, public opinion is a chaos of
superstition, misinformation and prejudice.**

(Gore Vidal, 'Sex and the Law', 1965)

The world's first opinion poll was held in Delaware in July
1824 by the *Harrisburg Pennsylvanian* newspaper to predict
voters' intentions in the presidential election later that
year. The main candidates were Andrew Jackson and John
Quincy Adams; Jackson emerged as a clear winner in the
poll, securing some 70% of the vote.

Jackson did indeed win in Pennsylvania, even surpassing
that 70% figure, and he did gain a narrow majority in the
national vote, but he fell well short of gaining an overall
majority in electoral votes. Under the provisions of the
Twelfth Amendment to the US Constitution, the decision
of who should become president passed to the House of

Representatives and they chose John Quincy Adams.

So you could say that opinion polls got off to an unconvincing start, and even though polling techniques have become increasingly sophisticated over the years, their predictions can still fall wide of the mark. The UK general elections of 2017 and 2019 are cases in point. In 2017, eight out of nine major opinion polls conducted one or two days before the election predicted Conservative overall majorities of between 28 and 92 seats. The remaining poll predicted a hung parliament with Theresa May's Conservatives 24 seats short of an overall majority. The result of the general election was a hung parliament with the Conservatives eight seats short.

Two years later, the polls gave a good prediction of the percentage votes each party would receive nationwide, but were less successful in detecting local differences and predicting the number of seats each party would win. All the main polls taken in the last few days before the 12 December 2019 election agreed that the Conservatives would probably win an overall majority, but they all hedged their bets saying that the result was too close to rule out a hung parliament. In fact, the predictions of the main polls for the Conservative majority ranged from 24 to 52 seats; in the election itself, the majority was 80. So what went right and what went wrong?

At the heart of any poll or survey lies the size and nature of the sample of respondents. The earliest polls, such as that conducted in Delaware in 1824, just picked people at random, presumably hoping that a random selection of people had a good chance of reflecting the general population. On the other hand, such a sample is inherently liable to be biased as it consists only of people who are willing to communicate their intentions to pollsters. It could be that the liberal attitudes of Democrats would make them more likely to respond than Republicans, who might be more conservative and suspicious of newfangled polling. To avoid, or at least minimize, such bias, ever more sophisticated techniques evolved under the general name of stratified sampling.

The basic idea is to break down the general population into various layers or strata by sex, age, employment status, etc., then select a sample of people for the poll which accurately reflects the distribution of all these factors in the population as a whole.

Around 2005, Swindon had the reputation of being the most average town in Britain, which made it a favourite among marketing organizations to test a new product. If something went down well in Swindon, the assumption was that it would thrive in the country as a whole. More recently, Swindon has become more affluent and, in 2017, Worcester was reported to have replaced Swindon as the most average town.

Some polling organizations rely on a panel carefully selected to match the average of the whole population they want to measure. Others are less precise about the selection of their panel but will gather data about them in

Stratified Sampling

Anyone wanting to conduct an experiment or survey on a random sample of a target population is liable to run into the same problem: there is no such thing as a random sample. Unless you test everyone, which is usually both impossible and prohibitively expensive, a selection method must be applied. Stopping people in the street restricts the sample to people walking the streets or going shopping at whatever times of day the data collectors are working. Relying on volunteers, or Internet users, or people who don't mind wasting time on the phone, or any other method of finding respondents, will run into similar problems.

Stratified sampling is the best way to minimize such effects; it works by dividing the desired population into various strata, for example by age, education, ethnic group, sex, income or other characteristic, and selecting respondents to provide a sample that reflects the make-up of the population. If such a precise composition is too difficult, then any sample can be used, but with the responses of the various strata weighted according to their relative numbers in the general population.

order to weight their response patterns to reflect differences between the average respondent in their sample and the overall national average. As a simple example, if the panel are 60% female and 40% male, we can multiply the male response pattern by 1.5 to restore the male–female balance and predict what the nation as a whole think.

Such ideas of stratified sampling were recently enhanced by the introduction of a technique with the very impressive name of multilevel regression with poststratification, or MRP for short.

This is not as complicated as it sounds. Regression is a clever statistical tool to find out how various factors interact to produce a combined effect. For example, obesity, lack of exercise, old age, sugar consumption and blood glucose level may all be seen as factors that can influence the onset of type 2 diabetes, but they also influence each other. Lack of exercise may increase the chance of obesity for instance, and increasing age may correlate with any of the other factors. A regression analysis can not only determine which are the primary factors and which are secondary, but also provide a formula to calculate, from any combination of the factors taken into consideration, the chance of any predicted outcome, whether that outcome is contracting diabetes or voting for Boris Johnson in the December 2019 general election.

Back on MRP, 'multilevel regression' refers to a regression analysis on the factors that might influence an individual's voting behaviour, while 'poststratification' is the way the results of such an analysis are applied to obtain predictions for each constituency. The manner in which this is done is part of the latest wrinkle in opinion polling. The obvious way to do this, of course, is to conduct polls in each region, but that is hugely laborious and expensive. Instead, the nationwide vote-predicting formula obtained from the regression analysis is used with data from the latest available census to produce a result for each constituency. The analysis tells us how people are likely to vote depending on their sex, age, education, employment status, marital status and whatever else was put into it, the census provides the figures we need to determine the composition of the electorates in different constituencies and so predict how each constituency as a whole is likely to vote.

The pollsters had another advantage in the 2019 UK general election in that it came only two years after the previous such poll. Because of this, they could ask each person how they voted in 2017 and compare their answer with their expressed voting intentions in 2019. Are women more inclined to change their allegiances than men? Are the old more faithful to their political parties than the young? Are the overweight more steadfast than the thin? I do not know the answers to such questions, but you can be sure that the pollsters do and know how to use the results to predict which parliamentary seats are likely to change hands.

The modern technique of MRP should not be confused with the far older and simpler idea of regression to the mean, which is a different matter entirely.

Regression to the Mean

In 1886, the psychologist, statistician, pioneer geneticist and all-round Victorian genius Francis Galton wrote a paper 'Regression Towards Mediocrity in Hereditary Stature' in which he outlined an important discovery he had made while looking at tables of heights of people in many families. Tall parents, he found, did in general have tall children, but the offspring were not as tall as the parents. Equally, the children of short parents were indeed shorter than average, but again not as much so as their parents. If you calculated the amount by which the parents were taller than average, he found that their children would exceed the average by two-thirds of that amount. In other words, they inherited some of the tallness, but moved a little closer to the mean.

He suggested, incorrectly, that this was because children inherited characteristics not only from their parents but from all their ancestors further back in time. A child's genetic inheritance may, of course, consist of characteristics inherited by their parents from distant ancestors, but one cannot inherit ancestors' characteristics that one's parents did not have. Even when it was discovered that our genes come only from our mother and father, Galton's result still held. As the word 'mediocrity' developed negative connotations, the name given to Galton's discovery was changed to 'regression to the mean'.

The reasons behind it may be complex, and the inheritance of height is a good example. Modern studies suggest that a person's height is about 65%–80% due to inheritance and the remainder can be put down to environmental factors such as diet and lifestyle, which tend to be more randomly distributed among a population.

One consequence of this discovery is what has been termed the 'regression fallacy', which consists of an unjustified assertion of responsibility for a change that is nothing more than regression to the mean. Two examples illustrate this fallacy.

First, the introduction of speed cameras at accident black spots. When certain places on roads are identified as having suffered a large number of accidents, speed cameras are installed. When the number of accidents drops, the authorities praise themselves on their action, which reinforces their belief that speed cameras reduce accident rates. But, because cameras were installed at the places with most accidents, the rate was liable to go down on any future measurement because of the attribute of regression to the mean. To test the true effect of speed cameras, the authorities would need to install cameras at a control group of sites with low accident rates as well. Regression to the mean would suggest that accident rates at those sites with the lowest rates would normally go up in the future.

A second example could be deciding to punish students who perform least well in tests, believing this will improve their future performance. On the next test, these students do better, which is interpreted as 'proof' that the punishment regime works. But as the group who were punished were the worst-scoring, their results were liable to improve through regression to the mean next time in any case. In fact, a similar argument could be advocated against the practice of rewarding or otherwise praising students for excellent results. On the next test, they are likely to perform not quite so well, 'proving' that rewards and praise are counter-productive.

Before we leave regression, I should like to offer as an additional example a frivolous but statistically impeccable calculation I worked out several years ago when invited to watch the finals of a Miss Great Britain contest. I was not particularly interested in the event, but my invitation included dinner, which looked appetizing. As the contestants strutted their stuff on stage, I quickly turned my attention to the programme, which gave me details of the age, height and vital statistics of each young woman. This enabled me, when the results were announced, to perform a regression analysis on those figures to see if I could use them to predict the final result.

This analysis provides an example of an additional powerful benefit of regression analysis: the ability to create a formula to calculate the value of an item in which you are interested as a linear combination of the measured values affecting it. In case that sounds intimidating, I should mention that a 'linear combination' just means multiplying each of the factors by various numbers, then adding them all together.

In the particular case of the Miss Great Britain finalists, their ages were very similar so I just entered the figures for their height and bust, waist and hips measurements, together with a figure indicating their final placing, then clicked on a button, and the regression analysis came out with the following formula:

$$F = 30.7 - 0.59T + 0.03B - 0.73W + 0.99H$$

This gives F (feminine beauty) in terms of T (height in inches), B (bust measurement in inches), W (waist) and H (hips). The figures were scaled to lead to the highest value of around 12 for the winner, with lower values going down to 1 for those who were placed second to twelfth.

Remarkably, if you enter the figures for Marilyn Monroe as recorded by her dressmaker (height 65 inches, bust 35, waist 22, hips 35), the resulting figure for F is 11.99, which is about as close to perfect as one could hope for, according to the formula. (The source or date for these measurements, however, is not known.)

I was rather surprised to discover from the formula that the single most important factor was hip measurement (with a weighting of 0.99) followed by slim waist (the minus sign before 0.73 indicated that final placings correlated negatively with waist measurement, so slim waists were rated more highly). When I wrote a newspaper piece reporting this and giving the formula that resulted from my regression analysis to predict a contestant's final placing, my editor told me he was very happy to see it. As he put it, the analysis confirmed his previously stated view that 'bums are the new tits'.

Whether that reflects a nationwide shift in our criteria for assessing female attractiveness or whether it just shows the predilections of the judging panel is still an open question. As scientists are prone to say at the end of a paper detailing their results, further research is needed.

I have never had the necessary data to check the extent to which my formula for beauty accurately predicted the results of later contests, but I doubt that its efficacy would have continued. The opinions of beauty contest judges are, after all, liable to change over time just as those of voters in elections do.

Back on the relative sophistication of political polls, we must now ask why they were not more accurate in predicting the 2019 UK general election result. On this question, dedicated pollsters have a number of hurdles in their path.

1. However meticulous they may be in selecting their sample, they are limited by the fact that some people will never agree to talk to pollsters. Any sample is therefore inherently not fully representative.
2. Some people are secretly ashamed of their political views and will say anything to avoid divulging them. This is part of what psychologists call 'motivational distortion'.
3. People change their minds between the time of an opinion poll interview and the election itself.
4. Putting a cross in a box in the privacy of a polling booth is considered by many to be more serious than talking to a pollster and may therefore produce a different result.
5. The results of the polls themselves may influence a voter's intentions.

The last of these is the most intriguing. Consider the 2016 US presidential election when the vast majority of opinion polls predicted victory for Hillary Clinton. Is it possible that seeing their candidate ahead in the polls, many Clinton supporters thought it less urgent to vote? Would she perhaps have obtained more votes and won the election if her supporters had seen a Trump victory as possible or even likely? Such a polling result, however, might also have influenced potential Trump voters, so it is difficult to say what the overall effect would have been.

Motivational Distortion

Whether we are talking about how people fill in personality tests or how they respond to the questions of pollsters, one thing that can affect results is deliberate errors in responses. Some people try to respond in a way they think will impress whoever interprets the results; others respond in the way they think will make the psychologist or pollster happy. In either case this is called 'motivational distortion' – although another term for it is 'fibbing'.

Perhaps the biggest problem in personality testing is that the profile that emerges from an individual's responses is liable to be a mixture of what that person really is, what they think they are, and what they want the questioner to think they are. Only in extremely well-balanced people with good self-knowledge do those three profiles coincide. The psychologist or pollster's task is to separate what the respondent really is from the other two.

A similar factor may have influenced the result of the 2019 UK election. The polls on average forecast a Conservative majority of about 28 seats. Perhaps that was what the electorate wanted: a clear but not too large majority that could get things done but was not free to exercise dictatorial powers. Had the final polls predicted a majority of 80, it seems to me quite possible that many Conservative supporters would have thought that victory was in the bag and would therefore not have bothered to vote, particularly those who voted Conservative through fear of the alternative.

Politics, of course, is by no means the only excuse for an opinion poll. In the first four days of 2020, I saw in various media reports the following results from polls.

In the UK:

8.2% of people aged between 16 and 64 have never worked;

24% of people expect to buy more vegan/vegetarian food in the future;

26% of people say gin is their favourite Christmas alcoholic tipple;

34% of people say that religion is the basis of morality while 32% disagree, and the rest are undecided.

In the rest of the world:

8.6% of Canadians admitted they had driven when they thought they were over the legal alcohol limit in the past 12 months;

25% of Swedes think the time they spend on social media is 'meaningful';

32% of US adults want to have sex in a risky or unexpected place;

63.8% of Japanese women and 58.4% of Japanese men say that women should continue their careers after having children.

On reading such figures in newspapers, most people will just accept them and rush on to the next story. The spurious authority conveyed by such numbers has achieved its purpose, especially when figures are given to the nearest decimal place. Yet in all the above cases, the figure raises further unanswered questions.

How can 8.2% of people aged 16–64 in the UK have never worked when the UK unemployment rate given by the Office for National Statistics in December 2019 was only 3.8%? How can there be more than twice as many people who have never worked than the number who are currently unemployed? And what does this '16–64' figure hide? Does it include people who are still at school or university who would not generally be expected to have a job?

What should we read into the finding that 24% of people expect to buy more vegetables? Is that 24% of all people, including children, or 24% of adults who regularly go shopping? Also, that phrase 'expect to buy' may hide a great deal. What were the other questions asked? It's not difficult to get the answer you want if you lead up to it suggestively. Just start by asking people whether they eat their five-a-day, then whether they think they would be healthier if they did so, then whether they would therefore think about eating a more balanced diet, and finally whether, on thinking about it, they expect to buy more vegetables.

Who are all the gin drinkers? Was the survey conducted in an off-licence? Were responses restricted to spirits? And what's the point in asking people about morality and religion when 34% of them (that's what 'all the rest' works out at) presumably just stared blankly at the pollster when asked to rule on such a deep philosophical point? That 'undecided' may hide a multitude of complex opinions.

And what about the Canadians who drove under the influence but did not admit it, and what percentage of Swedes think responding to pollsters is meaningful? We might also ask how the 32% of Americans who want to have sex in a risky or unexpected place compare with the 34% of Britons who, according to a poll in a Sunday newspaper, 'admit to getting intimate on a beach'. I suspect that either the sample was highly unrepresentative or the respondents were exaggerating. Instead of 'admit to' perhaps 'claim' would have been more accurate.

That same survey, incidentally, reported that when asked to give the number of their sexual partners, the average for men was 15 while for women it was 10. One may conclude from this either that the men were prone to exaggerate, or the women understated the true figure, or a significant number of men had homosexual relationships. For if the entire population had given accurate answers, the averages for heterosexual relationships between men and women would necessarily have been the same.

Amusingly, a survey in 2018 in the USA reported that '41.3% of men and 32.6% of women admitted to lying about their sexual history. Overall, men were more likely to increase their number of sexual partners, whereas women were more likely to decrease it.' Sadly, they did not ask respondents whether they lied about lying about their sexual histories. I can only imagine how the questionnaire would have begun:

1. How many sexual partners have you had?
a) 1–10   b) 11–15   c) more than 15

2. Was your answer to question 1 a lie?
a) yes   b) no   c) I plead the Fifth Amendment.

3. Was your answer to question 2 another fib?
a) yes   b) no   c) See my previous answer.

The question of whether respondents are telling the truth is a major problem for researchers, but even if we disregard that, the interpretation of results and the way they are expressed can be a great source of misunderstandings.

For example, in March 2019, several UK newspapers reported a survey by GCHQ's National Cyber Security Centre which found that '22% of adults would turn to children aged over 16 for help creating online accounts'.

But hang on a moment. How many adults have children aged over 16 they can turn to? Does that 22% figure include adults who do not have such children or is it 22% of the adults with children over 16 or access to someone else's children over 16? And what does that 'would turn to' mean? Does it mean they say they would turn to such a child if one was available?

Surveys love quoting percentages, but we are rarely told who constituted the sample being surveyed, what precisely were the questions asked, whether the sample was self-selecting (as is the case with most online polls today), and whether the percentage figures quoted refer to all respondents or all respondents falling into a relevant but unidentified category. Such information is usually provided by those who performed the survey in the form of a 'margin of error' inherent in the results, but this is generally considered too technical and boring for most readers. Yet, without those details, it is often impossible to interpret the results with much confidence.

Margin of Error

On the day before the UK general election in 2019, an opinion poll was published putting Boris Johnson's Conservative Party on 41%, the Labour Party on 32% and the Liberal Democrats on 14%. These figures were for the entire country, but when figures were examined on a constituency basis, the poll predicted an overall Conservative majority of 28 seats compared with all the other parties put together. The pollsters, however, also declared that a hung parliament with no overall majority was within the margin of error of their poll. So what does this 'margin of error' actually mean?

Suppose we toss a coin 20 times and it comes up heads 10 times and tails 10 times. Then our prediction for the next batch of 20 tosses is 50% heads and 50% tails, but the real question is how confident we are about that prediction. Any quoted 'margin of error' comes with a specific confidence level, which is usually 95%. So the question is: if we believe that heads and tails are equally likely, what range of results includes 95% of all possibilities?

With coin tossing, we can work out exactly what is likely to happen. There are 2^{20} possible results of tossing a coin 20 times, which equates to 1,048,576 potential outcomes. Only one of these is all heads (0–20), and another is all tails (20–0). The 1–19 and 19–1 possibilities account for 20 results each, while 18–2, 17–3, 16–4 and 15–5 and the reverse of those give 380, 2,280, 9,690 and 31,008, respectively.

Adding together all those numbers gives a total of 43,400 sequences of 20 tosses that result in a 15–5 split or worse. That leaves 1,005,176 of our 2^{20} possible sequences between 14–6 and 6–14, and that is just over 95%. So we can say our prediction is 10 heads and 10 tails, with a margin of error of 4 either way. In the case of opinion polls, there are other matters that can affect our confidence and margins of error, such as the question of how much one can rely on the answers given to pollsters and whether people who respond to questionnaires reflect the entire population. Such things may strongly affect the accuracy of predictions but have nothing to do with mathematical confidence levels and margins of error.

CHAPTER 3

Risk and Behaviour

Our illogical attitudes towards risk

Human beings have invented the concept of 'risk' to help them understand and cope with the dangers and uncertainties of life. Although these dangers are real, there is no such thing as 'real risk' or 'objective risk'.

(Daniel Kahneman, *Thinking, Fast and Slow*, 2012)

Predicting human behaviour from surveys and statistics may present some difficulties but our inherent inconsistency and irrationality is an even greater problem, no matter how hard we try to be logical. This is true especially when we are faced with the task of assessing risks. People have different attitudes to risk-taking: some are ultra-cautious, some are courageous, some are reckless. But we are all, to some extent at least, irrational, as is confirmed by both experience and research.

Perceived risk causes people to alter their behaviour – often in an illogical manner. Do seat belts make people drive more recklessly? Did the threat of CJD at the time of mad cow disease cause more people to die of heart attacks through switching from lean beef to fatty lamb? Why are some people afraid of flying when statistics identify it as perhaps the safest mode of transport? Between 2002 and 2010, 266 people were injured and four died in accidents involving lifts in the UK. So why are so few people afraid of travelling in a lift? Per passenger mile, they are probably more dangerous than aeroplanes.

Fear of lifts is not totally unknown but, unlike even more unusual phobias, such as xanthophobia (fear of the colour yellow) or deipnophobia (fear of dining or dinner conversations), it does not seem to have a specific name. Curiously, sufferers are frequently classified as either claustrophobic (fear of confined spaces) or agoraphobic (fear of open spaces), though these are often seen as opposites. Agoraphobia, however, can include a general fear of places or situations that might cause someone to panic or feel trapped, and that may include lifts.

When the first escalator on the London Underground was installed in 1911, a one-legged man known as 'Bumper' Harris was employed on the day it opened, 4 October, to ride up and down the escalator to show that it was safe enough for a man with a wooden leg to travel on. This was supposed to quell people's fears of the moving staircase, but may have had the opposite effect if people wondered whether an escalator catastrophe was the reason for his wooden leg.

Most of us, however, can laugh in the face of a lift or escalator, given that we encounter far greater potential hazards every day. Indeed, if you are an adult aged between 35 and 54 in Europe or the USA, there is roughly a one-in-400 chance that you will suffer a fatal accident of some sort within a year. In the UK, people bought lottery tickets for 21 years in the hope of scooping the jackpot, with a one-in-14 million chance, when there was a one-in-400 chance that they wouldn't even survive the year. There's a better chance of being killed in the next 20 minutes than winning the lottery. Then in 2015, the number of lottery balls was increased from 49 to 59, making a jackpot win even less likely. And still people went on buying tickets.

Actually, I'm being a little unfair to lottery ticket purchasers. There are many other prizes smaller than the jackpot, but there is still no more than a one-in-50 chance of winning any prize at all. What makes it so popular, of course, is the massive lure of a huge, life-changing prize compared with the almost negligible cost of a ticket. All the same, it makes little sense as a gamble.

Perhaps we could consider lotteries as the modern equivalent of Pascal's Wager, which is an argument for living a godly life, put forward by seventeenth-century French philosopher Blaise Pascal. A rational person, he said, should seek to believe in God and live as though God exists because if this belief is wrong, then such a person will have lost only some pleasures, but if this belief is right, the person stands to make the infinite gain of eternity in

Heaven and avoid eternal torment in Hell. Similarly, the modern lottery offers a gain so large that it might as well be infinite, while risking a loss – the price of a ticket – so small that it is negligible.

In theory, we coolly assess risks and alter our behaviour in order to maximize expected profit and minimize risk, or at least to make decisions based on comparing the two. But, in practice, the evidence suggests that our behaviour is motivated by panic and innumeracy. Take, for example, our reaction to the discovery around 1996 of a handful of cases of a new variant of Creutzfeldt–Jakob disease (CJD) caused by eating the beef from cows infected with BSE (bovine spongiform encephalopathy).

The annual number of deaths in the UK from vCJD, as the new variant was called, peaked in 2000 at 28, which was less than one in 2 million of the population or one in 5,000 of the deaths that year. Yet many people considered the risk so high that they gave up beef and switched to lamb with its much higher attendant risk of causing a stroke or heart disease from its greater fat content.

Such an over-reaction to a statistically very small threat can be explained in several ways. First, there is the fear of the sudden realization that something we had thought to be safe might actually be dangerous. Even if that danger is very small, it is a bit of a shock to learn that it is there at all. Better the health risk of lamb, which we have always known about, than the new health risk associated with beef. Second, there is the natural tendency to flee from danger.

We're not running towards the sheep: we are running away from the cows. There is also a natural tendency to be more afraid in the face of something you have never heard of, such as CJD.

Perhaps the greatest role in this particular irrationality, however, is our difficulty in grasping the nature of large numbers. Just as we hope to win the lottery at the ludicrously long odds of 14 million to one, we also exaggerate the chance of contracting CJD, even when the figures tell us that there is only one death for every 45,000 cows eaten.

We will have more to say later about the way we can fail to understand large numbers, but for the moment let's look at some other mortality statistics. How worried are you at the prospect of drowning in the bath, or a fatal fall down stairs, or being murdered, or hit by a train, or falling victim to a bolt of lightning? The table shows the number of deaths from such causes in England and Wales for the years 2016, 2017 and 2018.

Cause of death	2016	2017	2018
Being hit by a vehicle when walking	89	70	64
Being hit by a train	0	1	1
Killed in a cycling accident	81	74	58
Falling down stairs	795	730	747
Drowning when having a bath	35	23	22
Falling into a bath and drowning	3	4	2
Being hit by lightning	0	2	0
Being a murder victim	596	712	622

In 2017, it was reported that US President Donald Trump suffered from bathmophobia – a fear of stairs and slopes, from the Greek *bathmos*, a step. Some call it climacophobia, from the Greek *climax*, a ladder or staircase. Neither word is to be found in the *Oxford English Dictionary* or in the American Psychiatric Association's *Diagnostic and Statistical Manual of Mental Disorders*, but both can be found in specific lists of phobias. Some, however, insist that climacophobia is a fear of staircases themselves, while bathmophobia is a fear of climbing them. Surprisingly, the above figures suggest that such fears may be more rational than the far more common brontophobia – fear of thunder and lightning.

Before leaving this morbid topic, I should mention another item listed among causes of death in England and Wales. This one did not occur at all in 2017 or 2018 but there was one case listed in the 2016 figures. It is described as 'Specific disorder of arithmetical skills'. I have no idea what this may consist of but perhaps I should append a health warning: Innumeracy can be fatal (but it's very unlikely).

A great deal of insight has been gained into the nature and extent of our erroneous mental efforts since the 1970s when the Israeli psychologists Daniel Kahneman and Amos Tversky conducted a pioneering series of experiments on human decision-making. Applications of their theories to investment strategy earned Kahneman the Nobel Prize for Economics in 2002. Sadly, Tversky had died in 1996 or he would surely have shared it.

One of their most perplexing and far-reaching discoveries was that a person's response to a rational question can depend as much on how the problem is presented as the logic behind it. The following three examples are good illustrations. All of them involve two slightly different ways of presenting a choice. In each case, the respondent is invited to choose between option A and option B.

1. The choice is between two treatments for an illness.

1A. Surgery: For every 100 people having surgery, 90 survive the operation; 68 are still alive a year later; 34 are alive at the end of five years.

1B. Radiation: For every 100 people having radiation, all survive the treatment; 77 are still alive a year later; 22 are alive at the end of five years.

When this choice was given to one group of respondents, Tversky and Kahneman report that only 18% chose the radiation option. A very different result was obtained with what seems like only a small change in formulation of the same choice:

1A. Surgery: For every 100 people having surgery, 10 die as a consequence of the operation; 32 die by the end of the first year; 66 have died by the end of five years.

1B. Radiation: For every 100 people having radiation, none die during the treatment; 23 die by the end of the first year; 78 have died by the end of five years.

Remarkably, when this choice was given to another group of respondents, the proportion that chose radiation rose to 44%.

By couching the figures in terms of survival, respondents were strongly influenced by the increased number still alive after five years, but when the death rates were given, respondents gave priority to the 100% survival rate for the treatment itself. Yet the figures are identical.

2. **This is a straight monetary gamble from the US economist Richard Thaler.**

2A. Students were told that they had won $30 and given the option to toss a coin. If it came down heads, they would get $39 but if it was tails, they would lose $9 and get only $21.

2B. Or they could decline the coin toss and just keep the $30.

Given this choice, 70% chose the coin toss.
A very similar choice was posed to another class:

2A. You can toss a coin to win either $39 on heads or $21 on tails.

2B. Or you can just opt for a straight $30 without a coin toss.

In this formulation, only 43% chose the coin toss. It seems that if you tell someone that they have won $30 they are willing to chance $9 of it to win even more, but if they have nothing to start with, they are more likely to settle for the certainty of having $30 rather than risking getting only $21. We shall see in Chapter 13 what capuchin monkeys make of a similar problem.

3. **Earning from an urn (an example by US psychologist Michael Birnbaum). You are required to choose between two urns filled with balls of various colours. Without seeing them, you pick one ball the colour of which determines how much money you receive. You know the values of the different colours and how many of each are in each urn.**

3A. Urn A contains 85 black balls which are worth $100 each, 10 white balls worth $50 each and 5 blue balls also worth $50 each.

3B. Urn B has 85 black balls worth $100 each, 10 yellow balls worth $100 each and 5 purple balls worth $7 each.

When presented with this choice, 63% of Birnbaum's subjects chose Urn B.

His next formulation, however, produced a very different result:

3A. Urn A has 85 black balls worth $100 each and 15 yellow balls worth $50 each.

3B. Urn B has 95 red balls worth $100 each and 5 white balls worth $7 each.

In this case, although all he has done is make things simpler by putting together balls of equal value, only 20% chose Urn B. This is difficult to explain, as the choice was essentially the same as before. Reducing the number of colours of ball seemed to make people more wary of the possibility of receiving only $7.

Kahneman and Tversky called such curious effects on decision-making 'failure of invariance' and developed a behavioural model to account for it, which they named Prospect Theory. This theory enabled them in many cases to predict successfully which of two alternative formulations of equivalent choices people would be more likely to opt for.

This was far from being Amos Tversky's only contribution to the understanding of human irrationality, for in 1969 he had already written a classic paper called 'Intransitivity of Preferences', which had profound implications for decision-making in general and economics in particular.

Transitivity is well known in mathematics as a property of relationships such as 'equals' or 'is greater than'. If $a = b$ and $b = c$, then $a = c$. Or if $a > b$ and $b > c$, then $a > c$. The question Tversky asked was whether someone can logically prefer a to b and prefer b to c, but still prefer c to a.

This question raised difficult questions for both psychology and economics. Both of these disciplines had, for two centuries or more, leaned heavily on the concept of utility. When people have to choose between various options, they are assumed to assign, either consciously or unconsciously, an overall value for the utility of each choice. This 'utility' will usually involve balancing various factors such as their price, the effort such a choice imposes, the time it takes and many other factors which may be difficult or even impossible to measure, but a notional figure is ascribed to the utility of each option and the selection is made by picking the highest scoring. If preferences are not transitive, however, this procedure goes round in circles: option *a* has higher utility than *b*, *b* beats *c*, but *c* beats *a*. Utility theory, if it is to work, assumes rationality in decision-making and requires not only the ability to assign utility values, but also that those values should be transitive.

My own acceptance of human intransitivity came many years ago when I was in a supermarket trying to decide which after-dinner mints to buy. Given the choice between Matchmakers (M) and After Eights (AE), I preferred the latter as I considered them tastier and only a little more expensive. But I preferred Bendicks Bittermints (B) to After Eights, for the same reasons: they were much tastier and again the difference in price was acceptable. But given the choice between Matchmakers and Bendicks, I would have chosen Matchmakers on the grounds that they were perfectly acceptable and I felt the large extra cost of the Bendicks mints was not justifiable. In other words, B > AE, AE > M, but M > B (where the > sign indicates preference).

I quickly realized that a similar argument explained why I was having such difficulty deciding which laptop computer to buy. At the bottom end of the range, a £199 machine did everything I needed, but for another £50 I could get something significantly faster, and for £50 on top of that, I could obtain a large increase in memory, and for another £50 I could get a computer that came with much more useful software. Moving up in increments of £50, I soon found myself looking at a state-of-the-art £2,000 computer, which was definitely better value than the £1,950 computer I had just looked at. At that stage, however, I asked myself why I was even looking at something so expensive when the £199 machine I had started with did everything I need.

I had not, at the time, read Tversky's paper on 'Instransitivity of Preferences', but was delighted when I first did so to find that he gave some very similar examples. Instead of my after-dinner mints, he mentioned the problem faced by a manager interviewing candidates for a job. The main criteria for assessing them were their scores on an intelligence test and their previous experience. The manager had some doubts about the reliability of intelligence tests, so if two candidates' IQ scores were close, he would choose the more experienced.

Candidate A had a higher IQ than Candidate B, but not by much, so the manager preferred B as he was more experienced. Similarly B scored a little higher than C, but again the difference was small enough to be ignored and C was preferred for his extra experience. But when the manager compared A with C, he felt obliged to prefer A as the difference in IQ scores was now sufficiently large not to be ignored.

On the same lines as my computer example, Tversky gave this tale about a man buying a car:

> His initial tendency is to buy the simplest model for $2,089. Nevertheless, when the salesman presents the optional accessories, he first decides to add power steering, which brings the price to $2,167, feeling that the price difference is relatively negligible. Then, following the same reasoning, he is willing to add $47 for a good car radio, and then an additional $64 for power brakes. By repeating this process several times, our consumer ends up with a $2,593 car, equipped with all the available accessories. At this point, however, he may prefer the simplest car over the fancy one, realizing that he is not willing to spend $504 for all the added features, although each one of them alone seemed worth purchasing.

Tversky went on to discuss the mathematical conditions needed for transitivity to apply to assigning utility values for decision-making, but this has remained a contentious matter among psychologists and economists for more than half a century since the paper appeared in the journal *Psychological Review* in 1969. Those who insist that utility values can be defined in a manner that guarantees transitivity have, I suspect, never tried to buy after-dinner mints.

On a much larger scale, people's general assessment of risk and their behavioural reaction to it are greatly affected by their assessment of the potential size of the danger something may pose. Our instant emotional reaction to a threat may easily be more extreme than any later analysis. Indeed, there is a strong argument that emotional over-reaction to possible danger has been a very positive factor in our evolution. 'If in doubt, run away' is a policy with great survival value. A good example of behavioural over-reaction was provided by the aftermath of the 9/11 terrorist attacks on the World Trade Center and the Pentagon in 2001.

More than a decade after the attack, a survey reported that 52% of Americans were worried or very worried about being a victim of terrorism. This was despite the fact that less than 0.01% of deaths in the USA are attributable to terrorism.

Shortly after the attacks, 40% of Americans said they were less willing to travel abroad and around a third said they were less likely to fly. According to a paper in the journal *Applied Economics* in 2009, 'Driving Fatalities After 9/11' (by G. Blalock, V. Kadiyali and D. H. Simon), 'After controlling for time trends, weather, road conditions and other factors, we find that travellers' responses to 9/11 resulted in 327 driving deaths per month in late 2001. Moreover, while the effect of 9/11 weakened over time, as many as 2,300 driving deaths may be attributable to the attacks.' By now, that strongly suggests that the number of traffic fatalities caused by 9/11-related fear of flying will have easily overtaken the 2,977 victims of the terrorist attacks.

Two months after the attack, a study reported that 12% of New Yorkers with personal experience of the attacks (having escaped from the Twin Towers, or known a victim, or helped with the rescue work) suffered post-traumatic stress disorder (PTSD) symptoms. Remarkably, 4.3% of Americans from outside New York showed a similar effect.

In the case of the last group, the effect on individuals was shown to correlate with the amount of TV and newspaper coverage they had seen at the time. The media, of course, know that disasters attract viewers and sell papers and their first reaction, whether it is a terrorist attack, an epidemic or a volcanic eruption, always seems to suggest a worst-case scenario as they indulge in a bidding war with rival news outlets offering larger and larger estimates of the numbers killed or likely to be killed. Early reports of 9/11 gave the possible numbers dead in the tens of thousands before the figure finally settled at fewer than 3,000.

Overestimation of danger is understandably most likely in the case of risks that have a low probability of happening, a high disaster potential if they do happen, and are outside the perceived control of the individual who may be affected. These are known as 'dread risks' and explain why so many more people are afraid of flying than driving a car. Plane crashes are usually fatal and the poor passenger can do nothing about them. At least with a road accident there is a good chance of surviving it, and people hope their quick reflexes may prevent it happening.

Our fear of dread risks is heightened by the way the media treat them. In 2019, the Global Change Data Lab at Oxford University produced a revealing table comparing the space given in newspapers to deaths from various causes with the frequencies, expressed as percentages of all deaths and total death-related news coverage, of each type of fatality in the USA and the number of times people searched for information about them on Google. The table gives a few of their findings.

Cause of death	Frequency	News reports	Google
Heart disease	30%	2.5%	2%
Cancer	30%	13.5%	37%
Road accidents	8%	1.9%	10.7%
Suicide	1.8%	10.6%	12.4%
Homicide	0.9%	22.8%	3.2%
Terrorism	<0.1%	35.6%	7.2%

As the report states, 'there is a disconnect between what we die from, and how much coverage these causes get in the media'. We might also say that we read the papers to learn about things that frighten us but if we want to know about something that is likely to affect us, we consult Google.

Another problem is encountered when we ask people how much they fear certain things. Their answers may reflect an unknown combination of various aspects: the magnitude of threat posed; the likelihood of its happening; the perceived unpleasantness of the suffering it can cause, both personally and widespread. Different people will combine these in different ways in a manner that is probably unknown even to them.

Paul Slovic, professor of psychology at the University of Oregon and president of Decision Research, a company which specializes in investigating decision-making under conditions of risk, wrote a highly influential paper in 1987 on 'Perception of Risk', which included a list of 30 activities that various groups of people had arranged according to perceived riskiness. The activities ranged from nuclear weapons and handguns to domestic appliances and vaccinations, and the ratings by groups that included the League of Women Voters, college students and experts showed some intriguing differences. While women voters and students both put nuclear weapons at the top of their lists, the experts only ranked it 20th. The experts, however, had X-rays in 7th place, which the students relegated to 17th and the women voters to 22nd. The biggest disagreements came with swimming, which just got into the experts' top ten, while the women voters ranked it 19th and the students ranked it in 30th place, less dangerous than anything else.

Slovic concluded that 'When experts judge risk, their responses correlate highly with technical estimates of annual fatalities ... Lay people's judgments of "risk" are related more to other hazard characteristics, for example, catastrophic potential, threat to future generations.' He also points out that 'Research shows that people judge the benefits from nuclear power to be quite small and the risks to be unacceptably great ... reflecting people's views that these risks are unknown, dread, uncontrollable, inequitable, catastrophic, and likely to affect future generations.'

To that list of unknown dreads, one might also add risks that are known but of which people have no personal experience, such as the Earth being hit by an asteroid of the size that wiped out the dinosaurs. According to one estimate, we can expect such a disastrous collision about once every 250,000 years killing at least 7.5 billion people, yet few people suffer from meteorophobia. (An asteroid is just a rock hurtling through space, orbiting the Sun; it is called a meteor if it enters the Earth's atmosphere and either burns up or hits us.)

Perhaps we just shrug off the threat from asteroids in the hope that by the time they happen we will have developed the scientific expertise to detect them in time and blast them out of the way. Or perhaps we reason that it's not so much of a risk anyway, as a death rate of 7.5 billion every 250,000 years only works out at an average of some 30,000 deaths a year, which is way behind traffic accidents.

As an apt way to sum up this chapter, Slovic quotes a comment made in 1979 by American political scientist Aaron Wildavsky: 'How extraordinary! The richest, longest lived, best protected, most resourceful civilization, with the highest degree of insight into its own technology, is on its way to becoming the most frightened.'

CHAPTER 4

The Mathematics of Sport

The randomness of winning and losing

We suggest that the best shooting strategy of penalty kicks may be to aim to the upper two corners. Proper training should help in reducing the possible miss rate of such kicks.

(Michael Bar-Eli and Ofer H. Azar, 'Penalty Kicks in Soccer: An Empirical Analysis of Shooting Strategies and Goalkeepers' Preferences', *Soccer and Society*, 2009)

In a letter to the highly respected journal *Nature* in 1996, Californian psychologist Nicholas Christenfeld made a profound observation: sports fans are attracted not only by their appreciation of grace, strength, brilliance, violence and excitement, but also by, as he put it, 'a finely calibrated mix of skill and chance in the outcome'.

Elaborating that need for delicate balance, he pointed out that 'Contests with too much chance are pointless as measures of relative ability. Those with too little chance in the mix provide no suspense.' In other words, spectators want the better player to win most of the time but love the thrill of an occasional upset.

These comments succinctly encapsulate the true nature of the appeal sports have to both spectators and participants. As we shall see, an understanding of the mathematics behind any sport gives us a proper perspective on the relative contributions of skill and luck on the result of a game, which can only add to the appreciation of a victory.

Christenfeld's calculations showed that his comments applied not only to the results of individual games but heavily influenced the manner in which different sports evolve their own structures to determine an entire season's results. The greater the role chance played in determining the result of a single game, the greater the number of games needed in a league season to determine the best team overall.

Taking seven different sports, from baseball (162 games in a season) to American football (16 games in a season), Christenfeld looked at how often teams were defeated by opponents who finished below them in the final league table. He showed that the chance that the best team would top the league at the end of the season was roughly equal for all sports. In other words, the structure of each sport has evolved to produce a number of games that provides the same balance between skill and chance and thus the same chance of the best team coming out on top at the season's end.

Suddenly everything became clear. Until I read Christenfeld's words, I found watching golf extremely boring. The players hit the ball a few times to get close to a hole, then tried to putt the ball into the hole. Sometimes they did so, sometimes they missed. The better players would, of course, succeed more often than weaker ones, but always I just shrugged. Now, however, I saw the sport in a more statistical perspective. What I was watching were fluctuations of a random variable. Just as results of a coin toss fluctuate randomly between heads and tails, or rolls of a die fluctuate between one and six, a golfer's putt fluctuates between success and failure. Only a series of coin tosses, dice rolls or golf putts can determine the extent to which the coin or die is biased or how good the golfer is.

Four rounds of golf (18 holes per round) comprise 72 trials, the results of which enable us to make a fair estimate of the player's skill. But even the best player might be beaten by an inferior one who has better luck with his putting or whose earlier shots leading up to the putt just happen to land nearer the hole, and when the best golfer is competing against a hundred or more other players, the chance that one of them will end up with a better score increases considerably.

Just as 72 tosses of a coin will give a good idea of whether it is biased or not, 72 golf holes will give a good idea of who the best player is, but in both cases we cannot be sure.

Let's look at football and the English Premier League as another example. In the 2018–19 season, the average number of goals scored by a team per game over the season was 1.41. Only two teams, Manchester City and Liverpool, managed greater than 2 with averages of 2.50 and 2.34, respectively. In the 90 minutes of a game (excluding the possibility of added time), that works out at one goal every 36 minutes for Manchester City and one every 38.46 minutes for Liverpool. Ninety minutes is not enough to expect a fair estimate of the true frequency of something that happens only once every 35 or 40 minutes. That is why we sometimes get upsets, which is what keeps the spectators excited.

The need to maintain spectator interest also helps to explain the curious rules of tennis. If we really wanted to determine which of two players was superior, we could alternate the right to serve, giving, let us say 100 serves to each player and see who has scored more points at the end. That would give a very high chance of determining the better player, but quite apart from being unexciting to watch, it would greatly reduce the chance factor. Instead, we divide the points into games and the games into sets, which turns a high-scoring contest in terms of points to a low-scoring game in terms of sets.

The statistics for the Wimbledon men's singles final in 2019 are a good example of how this changes the result: Federer won more points than Djokovic (218 to 204); Federer won more games than Djokovic (36 to 32); Djokovic won the match by three sets to two.

The scoring system for tennis could hardly be better designed to create such an anomaly. In top-class men's tennis, the server wins around two points out of every three. From that figure, one can calculate that a break of service will happen about once in seven games. That means that the most likely number of breaks in ten games is just one, which is enough to give the set to one player by a score of 6–4. The players may have played around 60 points, but it was one random service break that awarded all the points for that set to that one player.

Thanks mainly to two Dutch statisticians, tennis is in fact the game that has attracted the greatest mathematical analysis, and the results are hilarious.

Franc Klaassen and Jan R. Magnus were passionate Wimbledon-watchers working at the Center for Economic Research at Tilburg University when they had the idea for their 1996 discussion paper entitled 'Testing Some Common Tennis Hypotheses'. In 2014, they expanded their results to produce the book *Analyzing Wimbledon: The Power of Statistics*, where they explain their motivation with a revealing calculation:

> In a men's singles tennis match at Wimbledon, one point lasts about five seconds. One game lasts about six points, or 30 seconds. One set takes about ten games, or five minutes. And the match may take four sets, or 20 minutes. In reality, the match does not take 20 minutes but perhaps three hours. Only 10% of viewing time is taken up by actual play; the rest of the time must be filled by the commentator.

As they point out, commentators do not have an easy job but often rely on a repertoire of commonly accepted ideas: new balls provide an advantage; the player who serves first in a set has a better chance of winning it; the seventh game of a set is often the most crucial.

Klaassen and Magnus drew up a list of some 20 such received ideas and set themselves the task of using statistical methods to find out which were true. Their results led to the rejection of far more hypotheses than were accepted. New balls are not an advantage; the seventh game of a set is not particularly crucial; serving first is only an advantage in the first set; in later sets, the person serving first loses the set more often than not.

Actually this last point is easy to explain: you serve first in a set if you received serve in the final game of the previous set (other than when that game was a tie-break, which is comparatively rare). Since the receiver is less likely to win a game than the server, that means you are likely to have lost the previous set, which suggests that you may well be the weaker player and therefore more likely to lose this set as well.

Such statistical niceties may seem of little relevance to the tennis players themselves, but before we leave their sport, it is worth mentioning an interesting result from another Dutch statistician. In 1995, Lex Borghans of Maastricht made an interesting comment on the Wimbledon final in which Boris Becker lost to Pete Sampras, when he pointed out that Becker would have done better if he had served more double faults.

The statistics of the match were unarguable on that point: of his 140 first serves, Becker won 58 points from the 73 that went in but only 26 from the 67 that required a second serve. So 52% ($^{73}/_{140}$) of his first serves were valid and 48% were faults, but when the first serves went in, he won 79% ($^{58}/_{73}$) of the resulting points. For second serves, however, his success rate was only 39% ($^{26}/_{67}$).

If he had continued serving in aggressive first-serve style for his second serves, his double faults would have gone up from 15 to 32 (calculated by applying his 48% first-serve failure rate to his 67 second serves), but he would have won 41% (79% of 52%) of these second serves, which was better than the 39% his actual second serves achieved.

That is not the only aspect of serving that has attracted mathematical interest: the question of where the server should direct the ball and where the receiver should stand is modelled nicely by the branch of mathematics known as Game Theory.

Game Theory

Mathematics is usually associated with solving problems or making rational decisions, but what if one has an opponent who is trying to frustrate one's endeavours? That's where Game Theory comes in. It was developed by the Hungarian–American mathematician John von Neumann in the 1930s as a way of analysing two-player games in which the success or failure of a player's strategy depends not only on his own decisions but also on those of his opponent or opponents. Its classic application is to two-player, zero-sum games, which are essentially games in which one player wins and the other loses, or one player's gain is equal to the other player's loss.

The mathematics of Game Theory was extensively developed in the 1950s by John Nash, whose achievements were later depicted in the book and film *A Beautiful Mind*. His most significant contribution was to prove that any zero-sum game offered at least one set of optimal strategies for each player, in which no participant can gain by changing strategy if the strategies of the other player remain unchanged. Such a set of strategies became known as the 'Nash equilibrium' and this concept played a large part in the extension of Game Theory to fields far beyond their original application to simple two-player games and as diverse as economics, politics, biology and even medicine.

Between 1998 and 2001, two economists at the University of Arizona, Mark Walker and John Wooders, wrote a series of papers in which one important aspect of the tennis serve was compared with a simple coin-tossing type of game. In the game, the first player chooses heads or tails; the second player also chooses heads or tails, trying to predict which choice his opponent has made. If the second player guesses correctly, he wins, but if he is wrong, the first player wins.

The application to tennis concerns where the first player directs the serve: does he aim down the middle or will he go for an outswinger? The speed of the serve is such that the second player must first decide where to stand to receive serve, and also decide which direction to move in even before the ball has been struck, but the basic question is whether he expects the ball to go left or right.

Since there is one point at stake, which must go to one player or the other, this is a simple example of a zero-sum game: the amount won by one player equals the amount lost by the other. Around the end of the 1920s, John von Neumann proved that any such game, when played by people who know what they are doing, has an optimal strategy known as Minimax, by which the players will minimize their losses and maximize their gains whatever the opponent does. Walker and Wooders called their papers 'Minimax at Wimbledon' and set themselves the task of finding whether the top players were following von Neumann's rules, though it was very unlikely they knew they were doing so.

The theory predicted that if both players adopted optimal strategy, then the server would win the same proportion of points with down-the-middle serves as with outswingers. Furthermore the two types of serve would occur with equal frequency. Remarkably, both those predictions were found to be followed and were confirmed in later analyses by others.

The only manner in which the players were found not to follow optimal strategy was that Minimax theory advised them to vary their service type at random, but humans are notoriously bad at simulating random sequences even if they are top tennis players. When trying to replicate a random sequence of heads and tails, a human will switch between H and T far too often, with shorter sequences of all-H or all-T than would occur by chance. Similarly, top tennis players change the direction of their serve more often than they ought if they want it to look truly random.

The practice of penalty-taking in soccer is very similar and has attracted much attention by statisticians. The penalty-taker, just like the tennis server, must choose to aim left, right or centre, and high or low. The goalkeeper must try to anticipate the direction of the shot and leap left or right or stay where he is. The best goalkeepers store information on the penalty-takers they are likely to come up against in order to take decisions based on what they know, but again the most profitable strategy, for both penalty-taker and goalkeeper, is dictated by von Neumann's Minimax.

In a 2003 paper entitled 'Professionals Play Minimax', in the *Review of Economic Studies*, Ignacio Palacios-Huerta, who is an economics professor as well as being head of talent identification at Athletic Club of Bilbao professional soccer club, analysed 1,417 penalty kicks in professional leagues throughout Europe to test whether players were adopting the best mixed strategies. His conclusions were surprisingly complimentary to the footballers:

Empirical evidence from the behaviour of professional soccer players in penalty kicks provides substantial support for the two empirical implications that derive from the hypothesis that agents play according to equilibrium ... This one-shot, face-to-face play involves professional subjects. These subjects have had the time necessary to become proficient at generating random sequences and to develop the instincts and learn what is considered to be the correct way of playing two-person zero-sum games.

Other statisticians, however, have suggested that footballers still have much to learn from mathematical economists.

Let us move on from the statistics of penalty-taking to the mathematics of managers in the English Football League, a topic that has also aroused a great deal of interest. The question is: when should a team sack its manager?

In 1997, three British economists, Richard Audas, Stephen Dobson and John Goddard, published an intriguing analysis of football results between 1972 and 1993 under the title 'Team Performance and Managerial Change in the English Football League'. At first the effects of managerial change looked good. Comparing the results of the first 18 games under a new manager with the 18 previous results showed a significant improvement, but perhaps that was only to be expected. After all, a team only sacks its manager when it is doing badly, and improvement is always easier from a low base.

So their next step was to compare the results of teams who changed their manager with teams doing equally badly who had not changed manager. Surprisingly, the results showed that the teams staying loyal to their managers in hard times also improved results, and by a greater amount than those who sacked their manager. Their conclusion was that 'managerial change appears to have a harmful effect on team performance immediately following a managerial termination'.

Five years later, however, a new piece of research told a more complex story in a research paper by Chris Hope for the Judge Institute of Management entitled: 'When Should You Sack a Football Manager?' His analysis of six years of Premiership games showed that the life of a manager typically goes through several stages:

Inspire: when the team does well, inspired by its new manager.

Rebuild: when results slump as the composition of the team changes.

Age: when the rebuilt team takes proper shape and a consistently high standard is achieved.

Drop: when both manager and players lose their impetus and results decline.

Decay: when the manager fails to stem the dropping process and things get worse.

In order to answer the question posed in the title of his paper, Chris Hope performed an intensive computer analysis to answer questions concerning the length of the above processes and to come up with a formula to calculate when a team should ditch its manager.

The recipe he and his computer came up with advised limiting a new manager's honeymoon period to only eight games, after which performance should be assessed giving the last five games 47% of the total weighting. The manager should be sacked if his average point score per game drops below 0.74.

Let's return to golf, which is a fascinating game for the statistician, but first we need to introduce another measure of performance that is very relevant to many sports. We have already met the three types of average – mean, median and mode – but in describing a player's performance over an extended period, a measure of their variability is required. Two cricketers, for example, may both have a batting average of 50, but one may be a thoroughly reliable sort of chap whose scores predominantly cluster in the region between 40 and 60, while the other has them scattered all over the place between 0 and 100.

The statistical factor that is calculated to describe this is called standard deviation, which is described rather intimidatingly as the square root of the average of the squared differences from the mean – but stick with me, because it's really not all that complicated and it is terribly useful.

Standard Deviation

It may sound a little risqué, but standard deviation is far more standard than deviant. Basically it is a very useful measure of the spread of values of a variable: the range of values may be wide or narrow, and tall or squat. That is what standard deviation tells us.

Specifically, standard deviation is defined as the square root of the mean squared distance of all the values from the mean. That sounds intimidating, but it's really not that complicated. You first calculate the mean by adding up all the values and dividing by the number of them, then you find the distances from each value to the mean (by subtracting the mean from each value, or the value from the mean), then square each of those values and work out the average of those squares.

Taking squares of the distance rather than straight values gives a higher weighting to items a long way from the average, which turns out to be much more useful. The average of those squared distances is called the variance, and the standard deviation is the square root of the variance.

In the general case of a normal distribution (which we shall encounter shortly), we expect about 68% of the data to lie within one standard deviation of the mean, 95% to lie within two standard deviations and 99.7% to lie within three standard deviations. In the UK, the mean annual rainfall over the past century is about 930 mm with a standard deviation of 115 mm. This means that we would expect 700–1,160 mm of rain (that's within two standard deviations of the mean) in 95 years out of 100, and we would expect less than 585 mm or more than 1,275 mm rain only three years in every 1,000.

Let's look at an example of standard deviation:

Suppose two classes in a school, each of six pupils, take a test consisting of seven questions. The scores out of 7 for one class are 1, 2, 3, 4, 5, 6, while the other class get 0, 3, 3, 4, 4, 7. Both classes have a mean score of 3.5, but the second class shows both greater clumping around the mean and a greater divergence from it, with one pupil getting all the questions wrong and another all correct, neither of which happened with the first class. In other words, the second class showed a greater spread in its scores than the first. We need a measure that expresses this in numbers.

One idea would be to look at the distances of all the scores from the mean (the difference between the mean and each score, or between the score and the mean). That gives 2.5, 1.5, 0.5, 0.5, 1.5, 2.5 for the first class and 3.5, 0.5, 0.5, 0.5, 0.5, 3.5 for the second. Both of those give an average (mean) of

1.5, so that method also fails to discriminate between the two sets of scores.

If we square the distances, however, it gives greater weight to the extremes. The first group now has 6.25, 2.25, 0.25, 0.25, 2.25, 6.25 while the second group has 12.25, 0.25, 0.25, 0.25, 0.25, 12.25. Now the first group's average is 2.92 while the second group's is 4.25.

These are the averages of the squared differences from the mean – which, if you look back a few paragraphs, is a phrase from the definition of standard deviation. Now all we have to do is calculate the square roots of 2.92 and 4.25 to see the difference in standard deviation between the two groups. The first group has a standard deviation of 1.71 and the second a standard deviation of 2.06.

At last we have a figure that expresses the difference between the two groups, and the standard deviation turns out to be a very powerful factor in telling us how measures are spread throughout a population. Statisticians like to pretend that any bar chart of their measurements approximates to the so-called 'normal distribution' curve where the measurements are distributed more or less symmetrically, rising to a peak at the mean, then declining slowly to form a bell shape.

Bell Curve

The concept of standard deviation is closely linked to the idea of a bell-shaped curve. When you draw a graph of the measurements of anything, whether it is the heights of mountains, or the weight of people, or lengths of the borders of countries, or millimetres of rainfall, or students' test result scores, with the height of the graph varying according to the number of examples with any particular value, the results will most often take the shape of a bell, with small values on the left rising to the most frequent value in the middle, then falling off again as values get higher. The perfect bell curve is symmetrical, reaching the average value at its highest point.

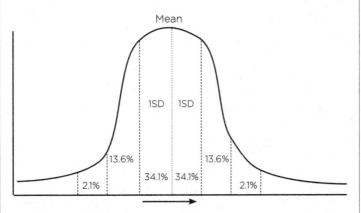

A normal distribution curve, plotting the values of whatever items one is measuring, increasing from left to right, against the number of items with each value, increasing from bottom to top. This is the characteristic bell curve, showing the percentages of the sample falling within one, two and three standard deviations (SD) of the mean.

This is the so-called normal distribution to which we hope that whatever we are measuring approximates, for the bell curve is the idealized model that lies at the heart of most simple statistics. Where things go wrong, the cause is often that the model is only an approximation of whatever distribution we are measuring. I suppose that is the main problem with mathematics: it describes smooth, calculable, predictable shapes but real life is messy. However, in general, the mathematics gives a pretty accurate picture of what is going on. In the case of many actual distributions approximated by a bell curve, we find that the bell is not symmetrical: its left and right sides behave differently (see diagram below). The average daily rainfall in the UK, for example, is around 2.5 mm but the figure for any particular day has been known to be more than 30 mm, while it is never, of course, below zero.

The graph therefore has a steep hump for the high point on the left and a long tail to the low point on the right. The asymmetry of a distribution curve is called 'skewness'. A long right-hand portion of the curve is called positive skew,

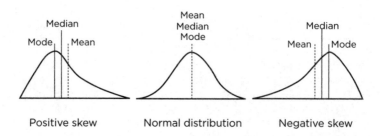

Positive skew   Normal distribution   Negative skew

a long left-hand portion of the curve is called negative skew. Adult male heights and weights conform well to the standard distribution; rainfall and personal incomes have strong positive skews; age at death has a strong negative skew.

This 'bell curve' in general works very well and the mean and standard deviation tell us a great deal. If the standard deviation is very low, the bell has a sombrero shape, but a high standard deviation produces something closer to a bowler hat. Sometimes, however, the fit with an 'ideal' normal distribution is less good and we need to be more careful in interpreting results. But when the full range of measurements fit the bell curve, the mathematics of the normal distribution can be very useful.

For example, one highly applicable thing that standard deviation tells us is the proportion of people in a population who are a certain distance above or below the mean. If the figures produce a curve that is close to the bell-shaped normal distribution, we know that 34.1% of people are within one standard deviation below the mean and another 34.1% are within one standard deviation above it. So in the UK, where the average height for adult males is 5 ft 9 in (175.3 cm) with a standard deviation of 2.9 inches (7.4 cm), we know that 68.2% of men are between 5 ft 6.1 in (167.9 cm) and 5 ft 11.9 in (182.6 cm) in height. The mean and standard deviation for women in the UK is 5 ft 4 in (162.6 cm) and 2.8 inches (7.1 cm), so 68.2% of women are between 5 ft 1.2 in (155.4 cm) and 5 ft 6.8 in (169.7 cm) in height.

The average weight in the UK is 84.8 kg (187.0 lb) for men and 72.4 kg (159.6 lb) for women with standard deviations of about 12 kg (26.5 lb) in both cases. Average heights for both men and women in the USA are about the same as in the UK, but their average weight is 88.8 kg for men (195.7 lb) and 75.4 kg (166.2 lb) for women.

Talking of heights and standard deviations, this is perhaps a good place to give a specific example of how the matters we have discussed may apply even when the bell curve is a little wonky.

The graph shows the heights of all 44 men who have held the title of US president, plotting height to the nearest two centimetres on the horizontal axis against the number of presidents of that height on the vertical. As you can see, the heights range from 164 cm (James Madison) to 194 cm (Abraham Lincoln).

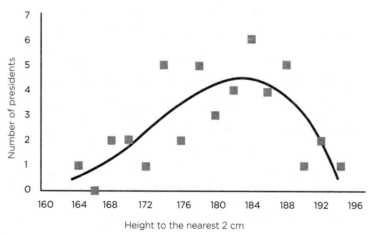

Heights of US presidents

The modal average (most common value) is 184 cm, which was the height of six presidents.

The mean average works out at 180.6 cm and the median is also around 181 cm, with 21 below it and 23 above it.

The standard deviation works out at 7.0, and we find that 34 presidents were within one standard deviation of the mean (174–188 cm), equating to 77.3% of the total of 44, which is not too far from the 68.2% predicted by a normal distribution.

The curve, which is the 'best fit' curve calculated by a regression analysis, clearly shows a distinct skew in the direction of taller presidents. This is not surprising in view of the fact that in 58% of US elections, the taller person won, and in 67% of them, the taller candidate secured a majority in the popular vote.

Surprisingly, similar statistical considerations to those we have been discussing (though not connected with height) also carry through to the sport of golf, where standard deviation plays a vital role.

In 2018, in a special sports issue of the journal *Chance*, statistics professor Stephen Stigler and his daughter Margaret, who has a degree in sports administration, wrote an intriguing article on 'Skill and Luck in Tournament Golf'. Their analysis is based on a very simple model for what determines a player's score for a round of golf:

Score = Par + Skill + Luck

'Par' is the score for the round that has been estimated to be the sum of the number of shots a strong player might be expected to take over all the holes.

'Skill' is a measure of the expertise of an individual player; in the above equation, the higher the numerical value assigned to 'Skill', the worse the player is.

'Luck' is a reflection of the fact that a player's scores on a particular hole will not always be the same, even under identical conditions. Again, good luck will be reflected by a low score.

The existence of luck is clearly demonstrated by the fact that in the four rounds of a golf tournament, all players may register four very different scores. If it were solely a matter of skill, only bad weather would have such an effect. Luck, if you like, could be defined as the supposed standard deviation in an individual's scores when playing the same course under the same conditions. By analysing such variations in scores, the Stiglers were able to estimate the relative contributions of Skill and Luck to the final result of a tournament.

Their conclusion was that the value of a player's standard deviation in a top tournament is around 2.85 strokes. The standard deviation, you will recall, is a measure of how much the measurements of something vary around the average; 34.1% of results will be within one standard deviation above the average, while another 34.1% will be within one standard deviation below it. So what the Stiglers were saying is that, in 68.2% of cases, a player's score was within 2.85 strokes of their average score on that course. In other words, the difference between good and bad luck was shown in a range spanning 5.7 strokes. They also concluded that the skill factor outweighs luck by about 4 to 1, but that still leaves a large role for luck.

The big question to answer, though, is what chance the best player has of winning the tournament. Naturally enough, that depends on the skill gap between the best player and the second best and, according to the Stiglers' calculations, if the top player has an average one-stroke advantage per round over number two, the probability of winning is about 31 in 100, or 31%, with another 16% chance of finishing second. If the best player has a two-stroke advantage, the probability of winning rises to 58%.

Under normal circumstances, however, statistics suggest that the best player has only a 0.4 stroke advantage over his nearest rival, and in that case the chance of winning reduces to 16%. In recent years, only Tiger Woods at his best was two strokes ahead of his nearest rivals.

These figures are hardly surprising. The best player's expected score may be lower than anyone else's, but his actual score will fluctuate by up to one standard deviation up or down in 68.2% of cases and by up to two standard deviations in another 27.2%. Meanwhile he is competing against 100 or more other players whose scores are also subject to similar standard deviations. It should hardly be surprising if one of those 100 ends up with a better score by pure chance.

I have often thought that the reason for the great popularity of games such as backgammon and poker is that they combine skill and luck in a highly seductive way. When players win, they can congratulate themselves on their skill, but when they lose they can put it down to bad luck. To some extent, we can now add golf to that list.

For all sports, however, the above analysis exposes a curious anomaly in our general attitude towards results. We like to think that consistency is something that should be admired and striven towards, but consistency is equivalent to a small standard deviation in a player's results. The golfer who makes a par score every time will always do well but only rarely win a tournament. Equally, a cricketer who scores between 40 and 60 runs every innings ends up with a magnificent average but much criticism for never making a century. We pretend to admire consistency but, as spectators, we reserve the greatest praise for those with high standard deviation whose bad results are forgotten when seen alongside the very good ones.

If a few very good results can create an exalted impression, what about a string of good results in a row? In all sports, there are periods in which a player seems to be in superlative form: a soccer player who seems to score goals in every game, or a cricketer who hits centuries in a string of matches, or a golfer who strikes a purple patch in which he holes every putt. But are these genuine instances of players hitting astoundingly good form, or are we reading too much into fluctuations of a random variable?

In 1985, the psychologists Amos Tversky, Thomas Gilovich and Robert Vallone drew attention to the 'Hot Hand Fallacy', which was the name they gave to a widely held belief that basketball players went through inspired phases in which almost any shot they made was likely to score. After confirming that players, spectators and coaches all generally subscribed to that belief, the authors performed a statistical analysis of a large number of basketball games to see if players really had such phases of success or whether it was just an example of our poor grasp of statistical fluctuations.

Their analysis considered every shot made and compared it with the previous shot by the same player. If a player went through good and bad patches, there would be a positive correlation between the results of successive shots with perhaps a series of successful shots even more likely to lead to another success. Their results, however, showed no correlation at all. The success of shots was determined merely by the expected fluctuations of probability.

Other researchers similarly investigated the results of players of other sports and generally came to the same conclusion: the distribution of a player's results over time is not an indication of changes in form during different periods but merely reflects statistical fluctuation.

For decades since the paper by Gilovich, Vallone and

Tversky, this has remained a contentious issue. Sports players and fans continued believing in 'hot hands' in basketball and purple patches in other games, while psychologists maintained firmly that it was just a delusion people fell into through an inability to grasp the implications of probability theory and a tendency to see patterns in random data. As Tversky himself once said, 'I've been in a thousand arguments over this topic, won them all, but convinced no one.'

Yet in 2014, three Harvard statisticians, Andrew Bocskocsky, John Ezekowitz and Carolyn Stein, challenged the findings of the original research, pointing out one possible factor that they maintained had not been properly considered: a player's reaction to his own 'hot hand'. Might it not be possible, they argued, that a player who has just made some successful shots is encouraged to attempt more difficult ones? They accordingly analysed a very large sample of basketball shots, taking into account the difficulty of the shot and its distance from the basket. Their analysis indeed confirmed that success led to players attempting more difficult shots, so while the original research had found no difference in performance after a series of successes, the Harvard trio drew a different conclusion. As Carolyn Stein put it: 'If your shooting percentage is staying the same [while the shots you attempt are getting more difficult], it almost seems like you must be becoming a better shooter.'

Even this new result, however, found the effect of the supposedly 'hot hand' was very small although statistically significant. At best it measured about a 1% improvement on performance. So the argument goes on.

One result that remains indisputable, however, is that people are not very good at statistical reasoning, especially where randomness is concerned. We have already met the tennis players whose serving and receiving strategy conforms very well to game theoretic recommendations, apart from their inability to change service direction randomly. Here are a few more examples that show similar points.

1. (From Kahneman and Tversky, 'Subjective Probability', *Cognitive Psychology*, 1972.) All families of six children in a city were surveyed. In 72 families the exact order of births of boys and girls was G B G B B G. What is your estimate of the number of families surveyed in which the exact order of births was B G B B B B?

When this question was posed to 92 students, 75 of them got it wrong, which is not surprising because it exposes three surprisingly common mistakes in most people's thought processes: a temptation to answer the question you thought ought to be asked instead of the one that was asked; a tendency to apply general principles to specific examples; and a delusion that patterns only occur rarely.

In this case, the correct answer (assuming girl babies are as likely as boy babies, which is more or less true) is 72. Since the sex of any newborn baby is independent of the sex of any that preceded it, any order of births is equally likely, from BBBBBB to GGGGGG via the more random-looking GBGBBG and BGBBBB. It is completely irrelevant that GBGBBG is three girls and three boys, which is more likely than one girl and five boys, because that was not what was asked. Most of the students asked fell into the trap and thought BGBBBB was considerably less likely.

2. I bump into an old friend and learn that he now has two children. 'Is either of them a girl?' I ask, 'because I happen to have in my bag a pink bonnet that would look great on a little girl.'

He accepts the gift gratefully saying it will indeed be appropriate. So I now know that at least one of his children is a girl. What is the probability that his other child is also a girl?'

If you reasoned that the other child is either a girl or a boy, with equal probability, so gave the answer 0.5, then you, like most people, are wrong. His two children are, before I learn anything about them, BB, BG, GB or GG, with equal probability. Learning that he has a daughter eliminates the BB possibility, so we are left with BG, GB and GG. So there is only a 1 in 3 chance that his other child is a girl.

3. **The Hungarian mathematician Pál Révész conducted a very revealing experiment in coin tossing with his students. He asked one group to toss a coin 200 times and record the results and asked another group not to bother with actually tossing a coin but just make up a sequence of 200 heads or tails. When they gave him the results, he was able, almost infallibly, to detect which were genuine and which were made up, simply by counting the longest sequences of tails in the results.**

In a series of 200 tosses of a fair coin, there is a 54% chance that somewhere in the sequence, there are seven or more tails in a row. The chance that the longest sequence is five tails or fewer is only 20%. The made-up sequences by Révész's students never contained long enough sequences.

Finally, while we are on the subject of probabilities, I ought to mention again the UK National Lottery. To judge from the number of Internet sites offering advice on which numbers are 'overdue' to be picked or which numbers are on a winning streak, having come up with great frequency in recent draws, there is a great deal of superstitious nonsense around.

In each draw, every number has the same chance of being drawn as every other number. The fact that a number has not appeared for a long time does not make it any more likely to be drawn in the near future.

The fact that a number has been drawn several times recently does not mean it is 'hot' and more likely to be picked again.

The fact that four consecutive numbers, 8, 9, 10, 11 appeared in the draw on 23 January 2016 does not indicate that the draw was fixed. Four consecutive numbers have appeared three times in the history of the UK Lottery, which is slightly below the number that would be expected by chance.

But here's a superstition that is definitely valid: odd numbers are more likely than even numbers. That is only because there are more of them, whether we are talking about the numbers from 1 to 49 in the early years of the lottery or 1 to 59 since 2015.

The only tip I have to offer, apart from not doing the lottery at all, is to avoid picking numbers that are well spaced. In January 1995, a record 133 people shared the jackpot when the balls numbered 7, 17, 23, 32, 38, 42 came up. The feature that made this selection so popular was their very characterlessness. They look random, and when a machine is picking random numbers, people expect the result to exhibit no clear pattern, so they keep their numbers well spaced out. When the draw features two or more consecutive numbers, there tend to be fewer jackpot winners.

Just bear in mind that it has been calculated that if you drive to a shop, buy a lottery ticket and drive home, you are more likely to die in a traffic accident than win the lottery, whatever numbers you pick.

CHAPTER 5

Saved You!

How governments try to look good

**Since problems are the only excuse for government,
solving them is out of the question.**

(P. J. O'Rourke, *On The Wealth of Nations*, 2007)

O'Rourke's delicious quip about the inadvisability of solving problems by governments is as astute as it is funny. He wrote that line in his very humorous and informative book about Adam Smith's 1776 *Wealth of Nations* treatise, offering it as an explanation of Smith's rambling style when he comes to talk about governmental problem-solving strategies.

There is, however, much more to be said about the way governments tackle (or fail to tackle) both problems and non-problems and it is all to do with their relentless pursuit of re-election.

When I was still in short trousers and attending primary school, there was a game we used to play whenever the opportunity presented itself, which happened only on rooftops or the edge of a cliff. When we saw another child on the edge of such a precipice looking downwards, the game was to sneak up behind them, grab them around the waist and shriek 'Saved you!'

There are essentially two ways governments may play their version of this game: exaggerating the threat posed by something, then claiming great credit for limiting its effect or even greater credit when nothing happens; or enacting a sledgehammer of reaction when a nutcracker would have done the job perfectly well. In many cases, a combination of exaggeration and over-reaction is employed to produce the greatest possible suggestion of government competence.

In 2007, in their book *Scared to Death*, Christopher Booker and Richard North gathered together accounts of numerous scares that had cost the UK huge sums in political over-reaction: 'The price we have paid for such panics has been immense; most notably the colossal financial costs arising from the means society has chosen to defend itself from these threats. Yet, again and again, we have seen how it eventually emerged that the fear was largely or wholly misplaced.'

They go on to say that many scares develop through an unholy alliance of politicians and journalists, the former seeing any threat, real or imaginary, as a chance to boost their reputation, and the latter knowing that there's nothing like a good scare story to gain readers or viewers, particularly if the newspaper or TV channel can claim a part in identifying the threat.

Booker and North then examine in detail a succession of over-reactions from the 1980s onwards to support their case. There was the great salmonella scare of 1988–9, which resulted in eggs being withdrawn from supermarket shelves despite there being no evidence that salmonella, which had been found in some chickens, could be transmitted to their eggs. There was the BSE/CJD crisis of 1996–9, first downplayed by politicians, then hyped in panic, leading to the slaughter of all Britain's cattle over 30 months old whether infected or not, which added up to some 5 million animals.

Among the non-food scary items, they include the Millennium Bug, passive smoking and asbestos, all of which saw claims made about their level of danger that were unsupported by the evidence. On the topic of passive smoking, Booker and North mention one study in 1989 which claimed that dogs living in the same house as a smoker had a 30% increased chance of contracting lung cancer. The author of the paper later admitted that lung cancer in dogs is extremely rare and his conclusions were based on the evidence of only one dog – a curiously inadequate piece of evidence on which to draw conclusions about the potentially harmful effect of passive smoking on humans.

The clamour to ban asbestos, which began in the 1980s, was similarly based on exaggerated fears. The term 'asbestos' covers a range of minerals, only a few of which pose health risks, and even those are a serious threat only when they are inhaled. Yet fear of the substance led to bans in many countries and a huge number of lawsuits in the USA, which cost the asbestos and insurance industries an estimated 200 billion dollars and brought about many bankruptcies.

Booker and North even cite global warming as an example of exaggerated claims of potential disaster, but we shall come to that in more detail later.

The UK, of course, is not the only country with such a history of over-reaction to scares. As just one example, in 2002 there was a cull of millions of pigs in Egypt during a potentially worldwide swine flu crisis. This happened despite there having been not a single case reported of swine flu among either pigs or people in Egypt.

In defence of politicians, we could say that it is their job to protect their people from disasters and to behave responsibly in that respect, so they must plan for a worst-case scenario. Epidemics should be prevented; the task of treating them when they have struck may be too great. Yet politicians should also examine all the evidence before rushing into action and shouting 'Saved you!'

In 2012, in the *Journal of Public Policy*, Moshe Maor, a professor of political science in Jerusalem, published an astute analysis of what he called 'Policy Over-Reaction'. Citing examples such as those given above, he identified several features they had in common: the willingness to predict rare events from weak evidence; the fact that negative events elicit strong responses and therefore have more weight given to them; and, perhaps above all, over-confidence among the decision-makers, who have too much faith in their own intuitions. As he says:

> Policy-makers believe they are more talented and competent than they actually are, have more control over the event at hand than they actually have, have more chance of success in solving the policy problems than they actually do, and perceive the information they possess as more precise than it actually is.

In a later paper on a similar theme, he added: 'One of the most common mistakes revolves around using the popularity of policy as an indirect measure of its worth.' Exaggerate the danger of something; take resolute action; and when it turns out to be far less dangerous after all, claim the credit and everyone will love you.

So is this all just a case of politicians wanting to claim undeserved credit or is there more to it than that? In 2013, the US online environmental magazine Grist published a story under the title: 'Science Confirms: Politics Wrecks Your Ability to do Math'. The report concerned an intriguing academic study by Yale law professor Dan Kahan which throws some light on what we have been discussing.

The study began by asking over 1,000 subjects some questions on their political views together with a series of questions to measure their mathematical reasoning ability. They were then given a tricky problem concerning the results of an invented scientific study.

The really clever part of the experimental design was that half the subjects were given a study supposedly assessing the effectiveness of a 'new cream for treating skin rashes', while the other half were given a study about the effectiveness of 'a law banning private citizens from carrying concealed handguns in public'. The data in the two versions were exactly the same, so exactly the same logic was required in either case to assess the results.

In the skin cream version, the figures gave the numbers of people who used the cream or didn't use the cream, and whether they got better or worse. Subjects had to decide whether using the cream was more likely to lead to improvement or deterioration.

In the handgun version, the figures were of cities that did or did not ban carrying concealed handguns, and whether the crime rate had increased or decreased. In this case, subjects had to decide whether a ban on guns was more likely to lead to a rise or fall in crime.

As a further embellishment, however, each version also came in two varieties, again using the same figures, but switched around to give the opposite correct answer.

The startling thing about the results was that they were very different in the two versions. In the skin cream version, responses were more or less as expected: the better a person had scored on the mathematical reasoning test, the more likely they were to get the right answer, and their political views did not affect that result. On the handgun question, however, politics was seen to play a big part. Democrats, who generally approve of banning concealed weapons, got the right answer when the figures supported that belief, but they gave the wrong answer when the figures supported the opposite view. Republicans, however, got it right when the figures suggested that the ban didn't work, but wrong when the figures supported the view that the ban did work.

Even more remarkably, the results showed that the more numerate somebody was, the more likely they were to let their political views adversely affect their conclusions. And that, perhaps, goes some way to explaining why supposedly bright people can hold such different views on emotive topics. Which brings us neatly on to the subject of the environment and global warming.

When the UN's Intergovernmental Panel on Climate Change (IPCC) brought out its first report on The Regional Impacts of Climate Change in 1998, the scientists behind it underlined the limitations of their methods with proper academic caution: 'Because of uncertainties regarding the sensitivities and adaptability of natural and social systems, the assessment of regional vulnerabilities is necessarily qualitative.' Where they cited quantitative estimates of the likely impacts of climate change, they stressed:

Such estimates are dependent upon the specific assumptions employed regarding future changes in climate... To interpret these estimates, it is important to bear in mind that uncertainties regarding the character, magnitude, and rates of future climate change remain. These uncertainties impose limitations on the ability of scientists to project impacts of climate change.

Most people, however, did not read the 517-page report but relied on summaries they read in the newspapers or saw on the television, and the media has always preferred stark warnings to caveats about limitations or uncertainties.

When a question is raised, we want an answer, not equivocation and more questions. So the public split into three camps over their interpretation of the original IPCC report and later follow-ups. At the two extreme ends, the most vociferous groups were shouting 'We're all going to die!' or 'This is just a bunch of doom-mongers trying to scare us', while a moderate group in the middle said little, thought that the predictions sounded a little alarmist, but believed that we should consider the fact that climate is very slow to change so we can probably ignore it.

The argument continued over the next couple of decades, and the science became more precise about the effect of carbon dioxide on temperatures. In 2018, the IPCC reported that CO_2 emissions would have to start falling 'well before 2030' and they would have to be on a path to fall by about 45% if global temperatures were to meet the goal set at the 2016 Paris Climate Agreement of no rise more than 1.5 degrees Celsius.

The newspapers, of course, in their customary sensationalist style, greeted this report with an announcement that we had only 12 years to save the planet. As I said, the media love stark warnings; the report and the exaggerated media accounts were latched on to by both sides of the climate debate in support of their beliefs.

At the World Economic Forum in Davos, Switzerland, in January 2020, the differing views were plainly heard from leading activists on both sides of the argument. The Swedish environmentalist teenager Greta Thunberg roundly criticized the lack of strong action by politicians, saying, 'We are facing a disaster of unspoken sufferings for enormous amounts of people,' while the US President Donald Trump said: 'We must reject the perennial prophets of doom. This is not a time for pessimism. This is a time for optimism.'

Perhaps we ought not to have been surprised at the result of that study in New York which demonstrated that people let their political views adversely affect their conclusions, and the more numerate someone is, the more likely that is to happen. As the global warming debate continues, the two people we are most inclined to listen to on the subject are a very political Swedish schoolgirl and a bombastic, not very well-educated US president.

As a recent related environmental issue has shown, a huge public response may shame politicians into taking action, but whether that action is commensurate with the problem is another matter.

In 2019, the UK government announced a ban on plastic drinking straws following a disturbing report by David

Attenborough on the vast amount of plastic pollution in the oceans that was threatening wildlife. Following that report, various surveys were conducted on people's attitudes to plastic in general and plastic straws in particular and it was reported that 80% of respondents wanted plastic straws banned. So the government did so, but their reasons for doing so seem far more to do with boosting their environmental credentials than making a significant positive contribution to the environment.

Channel 4 produced a very amusing Factcheck on the issue, which included the following points:

1. There seems no evidence to back up government estimates of how many plastic straws are used in Britain or how many end up in the oceans.
2. Straws make up only around 0.00002% of the weight of all global marine plastic pollution.
3. More than half of the global marine plastic pollution comes from China, Indonesia, Philippines, Vietnam and Sri Lanka. The combined total of every coastal EU country only comes 18th on the list of plastic polluters.

In any case, as plastic straws float, they are quite likely to be washed up again on beaches rather than contribute to ocean pollution.

So banning plastic straws is, quite literally, just a drop in the ocean. Responding in this manner to a strong result in an opinion poll is an excellent example of both over-reaction and under-reaction: the ban is an over-reaction to the poll result but a massive under-reaction to the underlying problem of plastic pollution.

In his excellent 1995 book *Risk*, John Adams, who knows more than anyone about risk in general and traffic risk in particular, mentions another striking example of misplaced reaction to risk: the accident black spot. Ponder this question: What is the best thing to do when statistics over a recent period of time identify a particular road intersection as having suffered a larger than average number of fatal accidents?

(a) Put up a sign saying 'accident black spot'.
(b) Institute prompt measures to widen the road, or improve visibility, or otherwise try to lessen the risk of accidents.
(c) Take no immediate action but keep monitoring the situation.

This is a far more complex decision than it may seem. There is a good deal of evidence that the first option, putting up a warning sign, makes people drive more carefully and reduces accidents, but surveys in Sweden have provided evidence that accidents may then increase at nearby junctions. Drivers slow down at the identified black spot, but then speed up again, just deferring the accident to the next junction.

For a similar reason, option (b) may also be ineffective. Road improvements may increase the sense of security of drivers, making them more willing to take risks. What matters is not so much the objective level of safety but a combination of real and perceived levels. Adams cites another study, also in Sweden, involving cars only some of which were fitted with studded tyres for winter conditions. The results showed that when the conditions were snowy or icy, drivers of cars fitted with winter tyres drove significantly faster than those in cars without them, but in clear and dry conditions no speed differences were found.

Option (c) seems the least attractive when action appears to be called for, but one argument in its favour is the concept of 'regression to the mean', which was discussed on page 39.

As we said earlier, in the case of traffic black spots, statistical studies suggest that if you monitor the number of accidents at a junction previously identified as a black spot, accident numbers in a later period are likely to drop, while numbers at spots identified as particularly safe are likely to go up, in both cases moving close to the average.

Adams quotes a lucid notional example suggested by another road safety expert, the Israeli–Canadian professor Ezra Hauer, to illustrate this procedure. Hauer envisaged 100 people all throwing a fair six-sided die. Around 16 or 17 (100 ÷ 6 = 16.7) of them would be expected to throw a six, so let's give each of those lucky people a glass of water to drink then ask them to throw again. This time, only two or three of them (17 ÷ 6 = 2.8) will throw another six. If we see throwing a six as something to avoid, then we could rejoice at the success of the 'water cure' – it worked perfectly in around five-sixths of cases.

It's a nice analogy: if you pick the road junctions in the worst sixth for accidents over a given period, then whatever you do is likely to produce an improvement, just as a glass of water supposedly stopped people throwing sixes.

Adams identified another related problem when looking into the question of whether it was a good idea to make the wearing of car seat belts mandatory. On the face of it, the matter was clear: the evidence definitely showed that wearing a seat belt reduced the chance of death for a driver or passenger in a car that was involved in a crash, but whether it reduced the chances of a crash were less clear and whether it made matters more or less safe for other road users was even more debatable.

Adams introduced the idea of 'risk compensation', which he explained as 'people modify their behaviour in response to perceived risk to their personal safety'. Apart from the obvious effect of people driving more carefully when they perceive danger, they are also likely to drive more recklessly when they think they are safe.

Figures on child road deaths in the UK over a long period provided strong evidence of something like this happening. As the number of cars on the roads increased, the objective level of danger to children increased, but this was more than compensated for by people's awareness of that danger. Between 1922 and 1986, the number of cars in England and Wales increased by a factor of 25, but the number of children under the age of 15 killed on the roads fell from 736 a year to 358. The death rate per motor vehicle dropped by 98%.

That reduction in death rate was perhaps due not only to road safety campaigns for children but also to risk compensation behaviour by their parents. In 1971, 80% of seven to eight-year-olds went to school on their own. By 1990, because of traffic fears and the great prominence given in newspapers to child abduction cases, that figure had dropped to 9%.

There is also strong evidence for risk compensation caused by seat-belt wearing. This is shown, for example, by comparing the road death figures for countries that had passed laws making seat belts compulsory with analogous figures for countries that had not done so. Collectively, those that had not passed such laws experienced a greater decrease in deaths over a given period than those that had passed laws. The evidence even suggested that the roads had become less safe for cyclists and pedestrians because more drivers were wearing seat belts and driving faster.

Fatal accident rates certainly decreased following the introduction of mandatory seat belts, but not by as much as they had been decreasing before, and not by as much as in comparable countries that did not bring in laws. But governments, of course, still claimed credit for the reduction.

Whatever the problem, if the original situation is bad enough, and some of that badness is due to statistical fluctuation, or can be eliminated by human awareness and sensible behaviour, then whatever you do – even nothing at all – will improve matters. Or, as the politicians might claim, 'Saved you!'

The kind of mathematical mayhem wrought by politicians may be summarized in five examples of the most common misuses of data, mathematics and statistics.

1. Citing inappropriate figures. Recent examples include the gain of £350 million per week gleefully emblazoned on the side of a campaign bus for the Leave side of the EU referendum in Britain. That figure wilfully ignored the £100 million rebate secured by the UK as well as any costs associated with leaving the EU. Also, in February 2020, in justification of a proposed points plan for UK immigration, the Home Secretary Priti Patel gave a figure of 'over 8 million' people in the British workforce who could fill the jobs that would no longer be performed by migrants. That 8 million figure included 2.3 million students, 2.1 million long-term sick, as well as many who are retired or looking after their family or home. Fewer than 1.9 million of the total were recorded as wanting a job.

2. Citing highly selective data chosen specifically to support proposed policies or justify existing policies. Examples include the use of police-recorded crime

data instead of more complete data that includes crimes not reported to the police, or hospital admissions data as a specific but incomplete measure of assaults. Another example of using selective data was seen in reports at the end of 2019 that life expectancy in the UK was falling. In certain areas such as Hartlepool, the latest figures did indeed indicate a slight reduction in life expectancy, but over the whole country figures had merely stalled, remaining constant for the past two years. When overall figures stay constant, one must expect some to fall and some to increase. Just pick the ones you want to justify your political case. Another example: in the UK in the year ending September 2019, on the one hand, the average custodial sentence length given in courts was the highest in the decade at 18.0 months, an increase of 0.7 months from the previous year. Hooray! We're being tough on crime. On the other hand, the total number of individuals formally dealt with by the criminal justice system in England and Wales fell by 1%. Boo! We're catching fewer criminals.

3. Choosing a manner of presenting statistics that creates the desired emotional reaction. Sometimes a figure presented one way will create a good impression while another presentation can look

depressing. Take infant mortality for example. On the one hand, according to the World Health Organization, infant mortality figures worldwide, giving the number of deaths before a baby's first birthday, are now at an all-time low of 29 per thousand births. On the other hand, stating that one baby in 34 still dies before it is one year old does not look like good news at all. If we say that 2,835 people in China died of Covid-19 virus in the seven weeks to the end of February 2020, it is very worrying, but expressing the loss as one-twelfth of the number of people killed on China's roads during the same period, or saying that the virus killed 0.0002% of the Chinese population makes it look far less worrying.

4. Claiming responsibility for something that is probably only a statistical blip or coincidence. Towards the end of August 1976, during a particularly dry summer in Britain, Denis Howell was made Minister for Drought and often referred to as 'Minister for Rain'. Days later, heavy rainfall caused widespread flooding, and he was made Minister of Floods. Apart from being encouraged by Prime Minister James Callaghan to do a rain dance for the nation, it is unclear what credit he may claim for the change in the weather, though he did reveal that he was helping water rationing by sharing baths with his wife. In general, it is in the

nature of politicians to claim credit when things go well and disclaim responsibility when things go badly.

Rather less fanciful examples frequently occur with crime figures. When rates of a particular crime are seen to rise, governments announce strategies designed to reduce the incidence. If the rate goes down, they proudly attribute the reduction to their strategy, though the true reason may be different. A good example was given by the figures for vehicle crime after 2002.

In that year, specific targets were set in a governmental Public Service Agreement to reduce high levels of thefts of or from vehicles. The British Crime Survey had estimated the number of such crimes in 1999 to be 2,956,000. By 2004, that figure had fallen to 1,886,000, representing a 36% decline, even exceeding the government target of 30%, which was naturally proclaimed as a resounding success.

According to an independent report, 'Ten Years of Criminal Justice Under Labour', from the Centre for Crime and Justice Studies published in 2007, however, vehicle crime had been on a steep downward trend for some years before the 2002 Public Service Agreement. Between 1995 and 1999, according to British Crime Survey figures, vehicle crime had fallen by 32%, from 4,318,000 to 2,956,000 and this was due mainly to car manufacturers introducing improvements in vehicle security. As the Centre for Crime and Justice Studies report stated: 'it was reasonable to assume that this trend would continue. Labour could therefore have felt fairly confident that this target would be hit, even if it had done absolutely nothing to make it happen.'

5. Comparing like with unlike. The detection and measurement of improvements and deterioration must involve comparing the latest figures with previous measurements, but methods of measurement also change, with the result that we may not be comparing like with like. Even something as basic as retail price inflation in the UK is measured according to changes in the cost of a basket of 'typical' household products, but the contents of the basket change over time. In 2019, 16 new items were added to the basket including popcorn, peanut butter, herbal tea, adult caps, a non-leather settee, an electric toothbrush and dog treats. These replaced certain items considered less typical of household shopping including a crockery set, soft drink in a staff restaurant, washing powder and envelopes. What effect these changes have on the overall figures is highly unclear.

Sometimes, comparing like with like becomes completely impossible because something is no longer being measured. In 2015, Prime Minister David

Cameron proudly announced that 'we will make British poverty history' and he did so by abolishing the 2010 Child Poverty Act and abandoning official measurements of poverty. So it could be said that he made poverty history, although, according to the independent Social Metrics Commission, in 2018 there were 14.3 million people living in poverty in the UK.

Finally, we should think about the response of various national governments to the problem of coronavirus, which we will look at in depth in the final chapter of this book.

The real problem for policy-makers was this: if you do nothing or take minimal preventative action and things turn out really badly, it will be politically disastrous not to have done enough. But the usual logic for exaggerating a threat applies: if nothing happens, or less severe bad effects are felt than predicted or even feared, you can say you gave a strong warning without which things could have been much worse. Or to put it more simply: 'Saved you!'

6	∞	9	3	8
2	7	=	1	%
4	¾	3	2	±
Σ	9	7	¼	3
¾	0	x	4	8

CHAPTER 6

Numbers Large and Small

How huge and tiny numbers confuse us

Human beings cannot comprehend very large or very small numbers. It would be useful for us to acknowledge that fact.

(Daniel Kahneman, 'This Much I Know', *Guardian*, 8 July 2012)

Towards the end of 2019, it was announced that the HS2 high-speed rail project linking the north and south of England would cost around £106 billion, which was more than three times the original estimate. What do you think on seeing such a figure?

(a) That sounds like an awful lot of money.

(b) How can they have made such a colossal miscalculation?

(c) Millions, billions, trillions – what's the difference anyway?

(d) How much is £106 billion anyway?

(e) All of the above.

Congratulations whatever you said, because any or all of the above answers could be considered correct. Yes, £106 billion is indeed a lot of money. Yes, underestimates are so common in securing government contracts that they have become quite normal. The trouble is that cancelling a hugely expensive contract tends to be so expensive that overspends are generally, if reluctantly, agreed. Inserting penalty clauses into such contracts is pointless as penalty clauses are liable to cause bankruptcy and the consequent loss of all the money ploughed into the venture. And, yes, even the richest and most numerate of us have problems in coping with large numbers when they are applied to money.

First, a bit of linguistic history. In the first edition of *Modern English Usage* in 1926, Henry Fowler wrote the following entry:

> billion, trillion, quadrillion, &c. It should be remembered that these words do not mean in American (which follows the French) use what they mean in British English. For us they mean the 2nd, 3rd, 4th, &c, powers of a million; i.e. a billion is a million millions, a trillion a million million millions, &c. For Americans they mean a thousand multiplied by itself twice, three times, four times, &c.; i.e. a billion is a thousand thousand thousands or a thousand millions, a trillion is a thousand thousand thousand thousands or a million millions, &c.

The word 'milliard' was also borrowed from the French and used in Britain occasionally to signify a thousand million, but it never really gained popular acceptance.

After this confusion persisted for around half a century, the British Chancellor of the Exchequer Denis Healey announced in 1975 that the Treasury would henceforth adopt the US billion for financial announcements, and a billion, in British English, has increasingly meant a thousand million ever since.

The word 'millionaire' incidentally was first used in French to describe the economist, gambler and dodgy financier John Law (1671–1729), who founded the Mississippi Company and made a fortune selling hugely overrated shares in it to the French, amassing more than 1 million livres in the process. The first reference in Britain was in *The Times* newspaper in 1795, which mentioned 'A certain Millionaire' whose 'absurd jealousies are the talk of all the world'. Sadly, the report did not tell us the name.

The first billionaire, according to newspapers in the US, was John D. Rockefeller, who was reported to have attained that level in September 1916 following a surge in the value of his Standard Oil shares. However, some say his fortune never exceeded $900,000,000. In any case, he fell well short of being a British billionaire as the exchange rate in 1916 was around $4.70 to the pound. Industrialist and carmaker Henry Ford is thought to have passed the billion-dollar mark sometime in the 1920s.

According to the 2019 Forbes Rich List, there are currently around 2,153 dollar billionaires in the world. Amazon founder Jeff Bezos is at the top of the list with an estimated net worth of $131 billion, but even he would not be able to afford the £106 billion estimate for HS2. At the current exchange rate of $1.30 to the pound, his fortune only just scrapes past £100 billion.

So how much is £106 billion, and can we judge whether it is likely to provide good value for the investment? Even when the original estimate of £32.7 billion was given in 2011, many doubts were expressed by independent bodies about whether it was worth the money. Just as millionaires must have seemed unimaginably rich a century ago, sums in billions of pounds are well outside most people's experience and are consequently difficult to grasp, so here are a few comparisons that may put the sum of £106 billion into perspective.

Items more than £106 billion:

£137 billion: total amount of revenue raised in VAT in the UK in 2019–20;

£130.1 billion: expected UK government health expenditure 2019–20.

Items less than £106 billion:

£29 billion: expected UK government defence expenditure 2019–20;

£63.5 billion: expected UK government education expenditure 2019–20;

£101.2 billion: expected state pension expenditure in 2019–20.

As another measure of value, £106 billion spread equally among all British households would cost each of them more than £3,500, and £106 billion is enough to pay the salaries of all the MPs at Westminster for 2,000 years at current prices.

Quite apart from the raw cost, if we look at the extent and complexity of a huge project such as HS2, it is not easy to judge whether something is or is not a fair price, but it is perhaps worth mentioning that the HS2 network involves 330 miles of track which, at a total price of £106 billion, works out at more than £5,000 per inch or £200 per millimetre). Is that good value?

Finally, we should mention that the value of all the notes and coins in circulation in the UK as of December 2019 was only £82.65 billion, so it is highly unlikely that anything costing £106 billion will ever be paid in cash. If we took 106 billion one-pound coins and piled them all up, they would not reach the Moon. If we changed them all into 50p pieces, however, they would reach the Moon, but only when it is closest to the Earth in its orbit.

Having seen how poor we can be at interpreting large amounts of money, we might ask whether we are any better at dealing with large numbers in general, which makes this a good place to bring in the Law of Large Numbers.

At its simplest, this basic law in probability theory tells us that when we are trying to measure the frequency with which something occurs, the more trials we conduct to get an idea of that frequency, the more accurate the final result will be.

Take the simple example of tossing a coin. Under normal circumstances, in the absence of any reason to believe otherwise, and assuming we have a fair coin, our expectation is that heads and tails both have a 50% chance of being the result on any particular toss. To check that assumption, we start tossing the coin and noting the results. How many tosses do we need to be reasonably confident that we can trust the overall result?

One toss of a fair coin has two equally likely results: H or T.

Two tosses have four possibilities: HH, HT, TH or TT.

Three tosses have eight possibilities: HHH, HHT, HTH, THH, HTT, THT, TTH, TTT, and the total number of possibilities doubles with each additional toss.

So eight tosses offer 256 equally possible results of which only one is all heads, one is all tails, eight result in seven heads and one tail, and another eight give seven tails and one head. And if we continue the calculations, we find that there are 28 ways for a 6H–2T split and another 28 for 6T–2H, with 56 for each of 5H–3T and 5T–3H and 70 for 4H–4T.

The total number of ways eight tosses can result in an 8–0 or 7–1 split either way (that's 8H, 8T, 7H–1T or 7T–1H) is $1 + 1 + 8 + 8$, which is 18 ways out of a total of 256. So tossing a fair coin eight times would result in unbalanced scores such as these more than 7% of the time. If we obtained one of these unbalanced results after eight tosses we might be tempted to think our coin was biased, but we need a larger sample of tosses to be more confident of what our results mean.

It turns out, if we do the calculations, that if we toss a coin 2,500 times and the results split equally between heads and tails, all we can say is that we are 95% sure that the chance of a head or tail on a single toss is between 48% and 52%. With 500 tosses, the range of confidence drops to between 45.6% and 54.4% and with just 100 tosses, a 50–50 result only allows us to claim with 95% confidence that the true balance is between 40% and 60%.

The Law of Large Numbers does not tell us that many tosses of a fair coin are increasingly likely to produce an exact 50–50 split (in fact just two tosses gives us the best chance of an equal split) but tells us more usefully that the greater number of times we toss the coin, the closer the measured result is likely to be to the true figure.

Now try this question:

Five tosses of a coin result in heads every time. What is your expectation for the next toss?

(a) Tails – because a tail is overdue.
(b) It's still 50–50. Past history does not affect the future.
(c) Heads, because from what we have seen so far, the coin may well be biased.

There is an argument for each of these answers being incorrect. If you chose (a), you have fallen for the gambler's fallacy, which is based on a misunderstanding of the Law of

Large Numbers. What that law says is that the longer your sequence of tosses of a fair coin, the closer the number of heads will come to 50%, which is not the same thing as saying that a 5–0 score will quickly be balanced by a lot of tails. If the next 50 tosses give an equal number of heads and tails, the score will be 30–25, which is definitely closer to 50–50 than 5–0. As Tversky and Kahneman put it:

> The heart of the gambler's fallacy is a misconception of the fairness of the laws of chance. The gambler feels that the fairness of the coin entitles him to expect that any deviation in one direction will soon be cancelled by a corresponding deviation in the other. Even the fairest of coins, however, given the limitations of its memory and moral sense, cannot be as fair as the gambler expects it to be.

Back on the question, if you chose answer (b) you could be accused of ignoring the evidence of the first five tosses. You may have started with a belief that the coin was fair, but after five heads in a row, your faith in that opinion may be beginning to shake a little. Five heads is, admittedly, too small a sample to conclude that the coin is biased, but at the very least you may well feel that if the coin is indeed biased, then there is a high probability that its preference is for heads.

Anyone choosing (c), however, is falling into the trap of another law, the Law of Small Numbers, which is the fallacy of jumping too quickly to a conclusion based on a small sample or expecting a small sample to behave the same way as a large sample. There are 32 possible outcomes of tossing a coin five times. The chance of five coin tosses all coming up heads is 1 in 32, and there is the same chance of five tails. So the chance of all five tosses being the same is 1 in 16, which is 6.25%. So the chance of the tosses not all being the same is 93.75%, which does not quite reach the 95% level a statistician normally requires to call a result 'significant'.

The quotation above about the gambler's fallacy came from a paper entitled 'Belief in the Law of Small Samples', in which the authors showed that even people who should know better may have unrealistic expectations in the behaviour of small samples. When asked what results they expected from hypothetical experiments on small samples, groups of psychologists consistently overestimated the chances that a study on a small group would confirm results obtained on larger groups.

That is why the direction of tennis players' serves are not as random as Game Theory would recommend and why people's simulations of sequences of coin tosses are not the same as the real thing (see Chapter 4 for these examples). We know that our serves and tosses ought, in the long run, to show a 50–50 distribution, but we make the mistake of following that rule in short sub-sequences of an overall longer sequence too.

To end this chapter on large and small numbers, here are a

few examples that may help you get some sort of feel for the numbers in our lives:

Most of us will live for more than two billion (2,000,000,000) seconds. Two billion seconds is just under 64 years.

The Equator is more than 1.5 billion inches long.

The world population is around 7.8 billion people.

The distance to the Moon is about 15 billion inches.

The average human brain has about 100 billion cells.

There are between 100 billion and 400 billion stars in the Milky Way.

Worldwide, around 300 billion emails are sent every day plus about 500 million tweets.

One light year, which is the distance light travels in a year, is about 5.9 trillion miles.

There are between 30 trillion and 40 trillion cells in the average human body.

The Sun is on average about 5.89 trillion inches away from the Earth.

It has been estimated that there are 10,000 trillion ants on Earth.

A kilometre is a thousand metres; a megametre,

gigametre and terametre are a million, billion and trillion metres respectively.

Beyond that we have petametre, exametre, zettametre and yottametre for a thousand trillion, a million trillion, a billion trillion and a trillion trillion (10^{24}) metres.

The Milky Way is about a zettametre wide; the observable universe is thought to be about 1,000 yottametres in diameter.

A centimetre and millimetre are a hundredth and thousandth of a metre; a micrometre, nanometre and picometre are a millionth, billionth and trillionth of a metre, respectively.

Beyond that we have femtometre and attometre for a thousand trillionth and a billion trillionth (10^{-18}) of a metre respectively.

The smallest known viruses are about 30 nanometres in length.

The diameter of an atom ranges from about 0.1 to 0.5 nanometre (1×10^{-10} m to 5×10^{-10} m).

A proton has a diameter of about one femtometre; the length of a quark is thought to be about an attometre ... and as far as we know, a quark is the smallest particle in the universe.

CHAPTER 7

The Insignificance of Significance

The misleading language of statisticians

**We must overcome the myth that if our treatment
of our subject matter is mathematical, it is therefore
precise and valid. Mathematics can serve to obscure
as well as reveal.**

(David Bakan, *On Method*, 1967)

Idly browsing through the UK newspapers one morning
in late January 2020, I came across the following attention-
grabbing headlines:

‘The average household contains discarded old
    technology worth £127’ (*Daily Telegraph*)

‘Technology plays a part in 72% of domestic abuse cases’
    (*The Times*)

‘China’s coronavirus DID come from bats’ (*Daily Mail*)

'Couples are experiencing an "intimacy deficit", says survey' (*Metro*)

As I suspected, however, when I looked more deeply into the stories, all those headlines were guilty of the usual journalistic exaggeration or obfuscation. Let's take them one at a time.

Discarded technology: as the report reminded us, mobile phones contain a tiny amount of many chemical elements, some so rare we have probably never heard of them. Quite apart from copper in the wiring, there is neodymium, terbium, tantalum and dozens of others, but it is the gold that the story wanted to excite us with. 'There is 100 times more gold in a tonne of mobile phones than in a tonne of gold ore,' they tell us, supposedly quoting from a UN report. Presumably this gold must contribute towards that £127 figure, but what do they mean by 'worth £127' anyway? Are they saying it's worth that much to the household that discarded it, or is that supposed to be the value of the gold, copper, neodymium, etc. after it has been meticulously extracted by experts?

If I bundle up my discarded mobiles and take them to a junk shop, will the proprietor believe me if I tell him they're worth 100 times their weight in gold ore? Naturally I wanted to check that fact in any case, but it seems rather dubious (though my calculations were made cumbersome by finding the available data in a mixture of metric and imperial units).

One pound of high-grade gold ore contains about 0.0005

oz of gold. There are 2,204.6 pounds in a metric tonne, so that works out at about 1.1 oz of gold in a tonne of ore, which is about 31 grams of gold. A few years ago, the US Geological Survey reported that a mobile phone contains 0.034 grams of gold. The most popular mobile phones now weigh on average about 160 grams, so there are 1,000,000/160 mobile phones in a tonne, which is 6,250. If each one of these contains 0.034 grams of gold as the geologists told us, then a tonne of mobile phones contains $0.034 \times 6,250$ = 212.5 g, which is well short of 100 times more than 31 g.

So the statement we started with looks like a considerable overestimate of the value of the gold in our phones. This confirms my innate feeling that if I bundled up all my discarded mobiles and took them to a junk shop, I'd be lucky to get a fiver, let alone £127.

What about technology playing a part in 72% of domestic abuse cases? The report explains that by 'technology', they are referring to the use of such things as hidden cameras and GPS tracking as well as social media harassment and various smart home devices, but what do they mean by 'plays a part' and what do they mean by 'cases'? Is every incident of domestic abuse a 'case' or only those that come to court? That 72% figure referred to cases handled by the Refuge charity, which is probably an unrepresentative sample of all domestic abuse cases. Does the 72% include all cases in which a technological device such as a mobile phone was mentioned (for example if a mobile phone was thrown at someone), or does it mean something more specific such as technology being used in a controlling/stalking way? Bearing all this in mind, giving a precise figure such as 72% implies an unjustified precision.

And DID (in block capitals) the coronavirus really come from bats? Well the headline said it DID, but the report that followed was less convincing. 'Scientists reveal that virus is 96% identical to the one found in [bats] ... This makes it the probable source – but this is yet to be confirmed.'

Oddly enough, 96% is also the figure often given for the similarity of human DNA to that of chimpanzees, though we sometimes see other estimates from 95% to 99% for that. It is a question of how the figure is worked out and whether it includes the 'silent mutations' that appear to have no effect.

The fact that coronavirus in humans is 96% similar to coronavirus in bats is not evidence in itself that we caught it from bats. Earlier reports had suggested that camels, snakes or civet cats may have been responsible, and since, at the time of the report, there was no direct evidence of any case of transmission of coronavirus from a bat to a human, even if it *did* come from bats, there is a high chance that it came via camels, snakes, civets or (the latest suspect) pangolins.

You will, I feel sure, want desperately to know to what extent people are experiencing an 'intimacy deficit', as found in the survey reported in the *Metro*. Among the rather dubious figures mentioned in this report, I was bemused to learn that 35% of Britons under 45 sleep closer to their phone than to their partner and even more astounded to see that 56% of all Britons crave more intimacy with their partners.

As is almost always the case with surveys, the report did not include the questionnaire given to respondents, so I do not know whether there was a precise definition of closeness given in the question. People tend to move around in their sleep, while mobile phones generally stay where they are put, often on the bedside table. So if two people are rolling around in their sleep, are we to take their average distance apart, or their furthest, or their nearest? And is that distance to be measured between their centres of gravity, or their two closest points?

The article tells us that the survey was of 1,000 adults, but does not say whether the sample was restricted to those who had partners to sleep with or included those who did not. The fact that the questions seemed always to mention partners suggests that single individuals were excluded, which makes the 56% figure for people wanting more intimacy even more disturbing. This means that at least 6% of people not only want more intimacy but are sleeping with someone who also wants more intimacy. There seems to be a problem with communication here. If they are too shy to talk, I suggest they grab the mobile phone, which is probably close by, and call the partner for an intimate discussion of the matter.

It is not just newspaper readers and journalists who are

often duped by numbers: even statisticians and psychologists have shown a disturbing trend of not quite understanding their own results. Indeed, a number of influential voices in recent years have suggested that one should beware of being fooled when a statistician or social scientist describes a result as 'significant'.

'Significant' does not necessarily mean the result is meaningful, still less that it should seriously alter your attitude or behaviour. Statistical significance has a precise meaning, very different from 'significance' as the word is generally understood. A significant result, for the statistician, is one that has probably not occurred by chance – and, even in that definition, the word 'probably' has a specific meaning. Yet it is a meaning that seems to have been forgotten by many who use it.

Significance

In 1925, the statistician Ronald Fisher introduced what quickly became a standard test of the extent to which the result of a test or trial, for example of a new drug, should be taken seriously, and for almost a century there has been a continuing debate about the value of Fisher's test of significance.

Fisher's idea was simple: you form a hypothesis and test it by performing an experiment that yields some data; then you apply a statistical technique that tells you the probability that your data occurred by chance, which therefore leads you to either confirm your hypothesis or reject it. There are three main things wrong with the way this approach is usually applied:

(1) Fisher's significance test involves his concept of the 'null hypothesis' (see separate box), which is actually the opposite of the hypothesis that the experiment is trying to confirm. An experiment is usually designed to confirm a difference between two groups being tested or surveyed. The null hypothesis is that there is in fact no difference between them, and Fisher's test tells us the probability (subject to certain assumptions) that the result of our experiment occurred by chance if the null hypothesis is correct.

This is the opposite of what we want to know. What we want is the probability that the null hypothesis is wrong (i.e. there *is* a difference between the two samples) given the result of our experiment. Essentially this is a telling example of a common confusion between the conditional probability of A given B and the conditional probability of B given A. Fisher knew of this problem but the misunderstanding is still frequently made, even by statisticians. Also, and perhaps more pertinently, nobody has yet produced a simple, generally applicable test of significance that is better than Fisher's and as easy to use.

(2) Using the word 'significance' is also a problem, fuelling a confusion between statistical significance and scientific or clinical significance. With a large sample size, a very small difference between groups may be statistically significant but the effect can be negligible when considering putting it into practice. Applying small differences between large groups to decisions involving individuals lies at the root of many prejudices.

(3) If you perform enough experiments, some of them are bound to produce statistically significant results. The American cartoonist Randall Munroe gives a delightful example of this in a strip cartoon of two stick people discussing scientific experiments to determine whether jelly beans cause acne. The scientist reports the results of his first experiment, which showed no link between jelly beans and acne. The stick girl tells him that she's heard that only a certain colour of jelly beans causes acne. The next 20 pictures show the scientist reporting no statistically significant links between purple, brown, pink, blue and 15 other colours of jelly beans and acne, but in the midst of these lurks one frame reporting a statistical link found between acne and green jelly

beans. The next frame shows a newspaper front page blazoned with the headline: 'Green Jelly Beans Linked to Acne' followed by the words '95% confidence' and 'Only 5% Chance of Coincidence'.

Out of any 20 independent experiments, you would expect one by pure chance to meet that 5% (one in 20) level of supposed significance. Green jelly beans are a delightful way of showing the striking manner in which this can be presented if you do not mention the other 19. As I said, if you perform enough experiments …

Most commonly, the result of a number of trials is held to be significant if the probability of its having occurred by chance is less than 1 in 20, or, as the experimenter will put it, the probability (p) of it being an accident is less than 0.05. A colleague once told me that he had been counting papers in academic psychology journals and found that 98% of them contained that solemn rubric '$p < 0.05$', which led him to conclude that around 3% had occurred by chance. The '$p < 0.05$' criterion tells us that we may expect the results of five out of 100 papers to have occurred by chance. That leaves 95 not occurring by chance, which is three less than the 98% count of papers carrying the '$p < 0.05$' seal of supposed approval. So the conclusion is that 3% did indeed occur by chance.

But it's that phrase 'by chance' that lies behind a great deal of misunderstanding and misinterpretation. In recent years, increasing numbers of statisticians and psychologists have drawn attention to the unjustified reliance that many in their professions have placed on statistical tests designed to demonstrate the significance of experimental results. For the chance being measured routinely by significance tests is often not what they think it is or claim it to be.

As the Californian mathematician and political scientist Jeff Gill put it in a paper entitled 'The Insignificance of Null Hypothesis Significance Testing' (*Political Research Quarterly*, 1999): 'Criticisms focus on the construction and interpretation of a procedure that has dominated the reporting of empirical results for over fifty years', a procedure he described as 'deeply flawed' and 'widely misunderstood'. He even supported his castigation of significance testing by quoting other eminent psychologists as describing it as 'one of the worst things that has ever happened in the history of psychology' or a display of 'mindlessness in the conduct of research'.

So what, you may be wondering, are they all doing wrong and what is this 'null hypothesis significance testing' anyway?

Null Hypothesis

When statisticians come up with ideas, or when they are commissioned to test someone else's idea, they test it out on relevant samples. That idea may be that the average Dutchman is taller than the average Briton, or that cats are cleverer than dogs, or that a particular coin does not have an equal likelihood of coming up heads or tails when tossed, or anything else that differentiates between two groups. Testing that idea then involves randomly selecting samples of the two groups and giving them a test of what you are trying to measure. Then you see if the Dutchmen and Britons, or cats and dogs, or coin tosses exhibit a difference in their scores.

Finally, you apply a statistical test to see whether the difference is marked enough to support the original idea. The 'null hypothesis' is that there is no difference between the two groups – that Dutchmen and Britons are much the same height, that cats and dogs are equally clever, and that the coin is fair.

Any psychological experiment is designed with the intention of showing that the null hypothesis is wrong, so we apply a significance test in the hope that it will confirm that belief. When the test comes up with a figure less than 0.05, we lift our arms in joy and report that the null hypothesis is highly unlikely to be correct – but that's not what the test is telling us at all. The trouble is, as mentioned above, that the significance tests we use do not tell us the chance of the null hypothesis being correct given the result of our tests, they tell us the chance of our result being obtained if the null hypothesis is correct.

This may sound like a petty semantic argument, but Gill puts it down to a simple logical fallacy. An elementary theorem in logic tells us that the statement 'A implies B' is equivalent to 'not-B implies not-A'. For example, saying 'All crows are black' is the same as saying 'All black things are not crows'. However, psychology and statistics deal with probabilities and tendencies, not statements of fact that are always true. Saying 'If A is true, then B is highly likely' is not the same as saying 'If B is false, then A is highly unlikely'.

For example: 'If a person is British, it is very unlikely they went to Oxford or Cambridge University, so if a person went to Oxford or Cambridge, they are probably not British.' Wrong!

More examples of what he called 'statistical buffoonery' were given by Charles Lambdin of Intel Corporation in 2012 in an article entitled 'Significance Tests as Sorcery' in the journal *Theory and Psychology*. 'Significance tests do not actually tell researchers what the overwhelming majority of them think they do,' he said before giving numerous examples of fallacious beliefs regarding significance. Such a test does not tell you the probability that the result will be replicated if the study is repeated; it does not tell you the probability that the null hypothesis is true; and it does not, as is most commonly believed, even tell you the probability that the result occurred by chance.

Echoing Gill's point about confusion between the probability of a hypothesis being correct given the experimental result and the probability an experimental result occurred given the truth of the hypothesis, Lambdin gives a neat example relating to abused children having nightmares.

When talking of 'conditional probability', which is the chance of something happening given the truth of something else, the standard notation for the probability of A (in this case abuse) given N (nightmares) is $P(A|N)$. 'Clearly', Lambdin writes, '$P(N|A)$ is not equal to $P(A|N)$.' The first is equal to the ratio of the number of abused children who have nightmares to the number of abused children. If all abused children have nightmares, then $P(N|A) = 1$. But $P(A|N)$ is equal to the ratio of the number of abused children who have nightmares to the number of children who have nightmares. It will only equal 1 if every child who has nightmares has been abused.

My own experience of conditional probability came many years ago when I felt the need to explain it to the members of a jury on which I was serving. An expert witness had presented DNA evidence, saying that it indicated a one-in-a-million chance of the similarity detected between the DNA of the defendant and that of the person who had committed the crime.

This sounded damning, but I pointed out that we had not been told why the defendant was arrested. If the DNA was the only thing linking him to the crime and he was arrested because his DNA was on a database, then we needed to know how many people were on the database. If it includes all 66 million people in the UK, then about 66 people would produce a one-in-a-million similarity. If, on the other hand, the police had other grounds for arresting him and the DNA was afterwards found to match, then the one-in-a-million chance would be highly significant (in the normal sense of the word).

I suggested to my fellow jury members that it was a bit like knowing the birthday of the criminal. If we have a suspect and he turns out to have the right birthday, that can add quite strongly to the evidence against him, but if the police started by rounding up everyone on their books with the right birthday, then the value of that piece of evidence is considerably reduced.

We sent a question to the judge to ask if we could be told what the grounds were for the defendant's initial arrest and he told us that we could not, which made it difficult to know the true significance of the evidence.

Such conditional probability lies at the heart of two mathematical misunderstandings known as the Prosecutor's Fallacy and the Defender's Fallacy.

The term 'Prosecutor's Fallacy' was coined by two

American psychologists in 1987 but since 1999 has been associated with a particularly notorious case in the UK. Sally Clark was convicted of the murder of her two babies, who had died aged two and aged three months. The defence was that the cause of death in both cases was Sudden Infant Death Syndrome (SIDS), but the prosecution produced an expert witness who testified that the chances of two babies in the same family dying of SIDS was vanishingly small.

According to his testimony, the chance of one baby dying of SIDS in the absence of known risk factors is about 1 in 8,543, so the chance of two consecutive such deaths is 1 in $8,543 \times 8,543$ which is about 1 in 73 million. Given the number of two-child families in the UK, he said this would therefore happen about once in 100 years.

That was his first mistake. The chance of two events both happening is only given by multiplying together their individual chances if the two events are independent. However, studies had already suggested that there was a genetic component to the occurrence of SIDS, making such a death more likely in a family that had already experienced it once. The chance of it occurring twice in the same family was much less than 1 in 73 million and instead of expecting it to happen once in 100 years, once in 18 months was considered a more likely figure.

His second mistake, however, was even more serious. Even if 1 in 73 million was the chance of Sally Clark suffering two consecutive cases of SIDS if she was innocent, it was far from being the chance that she was innocent given that the deaths had occurred, which was what the jury should have been considering. As one witness at Sally Clark's appeal pointed out, if we look at all the occurrences of double infant deaths, around one-third are caused by SIDS, one third by other rare medical causes and one third as a result of child abuse. That simple statistic would suggest that the chance of her innocence was more like 2 in 3 than 1 in 73 million. Indeed, given the rarity of infanticide by mothers, one could apply a similar method to the one presented at the original trial to obtain a figure of one in 2.152 billion for the chance of a mother killing her two children.

After one failed appeal in which the judges still seemed not to grasp the statistics of the case, a second appeal was successful but more because the prosecution was found to have withheld pertinent medical evidence from the defence than for reasons of misapplied statistics.

This basic error of thinking that $P(A|B)$, the chance of A given B, is the same as the chance of B given A, $P(B|A)$, has also been used by defence lawyers. For example, when the defendant's DNA is matched to the crime scene, and the chance of such a match has been calculated to be 1 in 10 million, the defence may point out that there are over 7.5 billion people in the world, so around 750 of them would have DNA that matched, so the chance that the defendant is the criminal is only 1 in 750.

The basic idea behind all this is Bayes' Theorem, named after the early eighteenth-century English philosopher and Presbyterian minister Thomas Bayes (c.1701–1761), who was a statistician even before the word 'statistics' was first used (the *Oxford English Dictionary* gives its first citation as 1770). Bayes' Theorem is stated as follows:

$$P(A|B) = \frac{P(B|A)P(A)}{P(B)}$$

To put it another way, the ratio of the probability of (*A* given *B*) to the probability of (*B* given *A*) is exactly the same as the ratio of the probability of *A* to the probability of *B*:

$$\frac{P(A|B)}{P(B|A)} = \frac{P(A)}{P(B)}$$

The bigger the difference between the chances of *A* and *B* occurring, the greater the magnitude of making the mistake of thinking that (*A* given *B*) and (*B* given *A*) are equally likely.

Bayes' Theorem

This theorem lies at the heart of many statistical misunderstandings. A typical example is a test for a disease. Suppose there is a disease that is known to affect 1 in 10,000 people and we have a test that is correct 99% of the time. What is the chance that a person testing positive actually has the disease?

Suppose we test a million people. Then we would expect about 100 of them to have the disease. Of those 100, the number testing positive will be 99. But there are also 999,900 who don't have the disease, and 1% of them, which is another 9,999, will also (wrongly) test positive. So the total number testing positive will be 10,098 of which only 99 will have the disease. So the chance that a person testing positive has the disease is only $\frac{99}{10,098}$, which is less than 1%.

What we are interested in here is conditional probability: the probability that A is true given B, or $P(A|B)$ as statisticians express it, and that's what Bayes' Theorem is all about, relating $P(A|B)$ to $P(B|A)$:

$$P(A|B) = \frac{P(B|A)P(A)}{P(B)}$$

The simplest way to see what this is getting at and to prove it is with the aid of a picture called a Venn diagram:

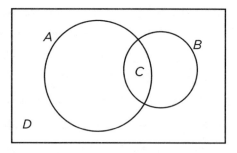

The rectangle D represents the whole population, the circles A and B are the parts in which we are interested and *C* is their intersection, comprising everything that is in both *A* and *B*. Let's call the number of data/ people in *A*, *B*, *C* and *D* respectively *a*, *b*, *c*, *d*. Then

$P(A) = a/d$

$P(B) = b/d$

$P(A|B) = c/b$

$P(B|A) = c/a$

It is relatively straightforward to verify that

$$\frac{c}{b} = \frac{(c/a)(a/d)}{(b/d)}$$

which is what Bayes' Theorem states.

This misunderstanding of statistical significance, however, is only one aspect of the problem in general understanding of statistics. Perhaps an even greater difficulty is the confusion between the way statisticians use the word 'significance' and the way we use it in the real world. The trouble is that a very small difference between two groups may be mathematically significant without even coming close to being significant in the normal sense. All you need to make an insignificant difference significant is a large sample.

Take coin tossing, for example. If you toss a coin 20 times, you need the difference in the score between heads and tails to be at least 15–5 to judge that the coin might be biased. Anything closer than 15–5 (i.e. scores 10–10 to 14–6) falls within the 95% of most likely outcomes. But 15–5 is a 75%–25% split between heads and tails, which is huge. On the other hand, if we toss the coin 10,000 times, we may expect between 4,900 and 5,100 heads (and the corresponding 5,100 and 4,900 tails) with 95% confidence. That's only a 49%–51% split.

One might say that a coin coming up heads only 4,900 times in 10,000 tosses hardly matters a toss to anyone using it to decide something, but the difference is significant to a statistician. With a small sample, we needed a massive difference for a statistically significant result; with a large sample only a tiny one was enough. As we have already pointed out in the context of coin-tossing trials, most surveys and experiments are conducted with as large samples as possible specifically in order to get statistically significant results.

Especially in the field of personality testing, results obtained from large groups are then used to make judgements about individuals. This may produce marginally improved results in selection processes, but is essentially only slightly better than tossing a fair coin. Small but significant measured differences between men and women, or white and black people, or even blondes and brunettes can be used to provide pseudo-scientific justification for decisions based on prejudices.

Research has shown, for example, that blonde waitresses on average earn higher tips in restaurants than brunettes. This is not considered a legitimate reason for selecting only blondes for waitress jobs in a restaurant.

On the other hand, some studies have supported the conclusion that extraverts make better investment bankers than introverts, possibly because their social skills are an advantage in dealing with clients or customers. Such findings as this may be used in the application of personality tests for selection purposes. The effect of hair colour on bankers' performance has, as far as I know, not been investigated.

CHAPTER 8

Cause and Effect

Common logical confusions

¾	0	x	4	8
3				+
6	4			9
¼	0			7
5	%			3
0				½

**There is no more dangerous error than confounding
consequence with cause.**

(Friedrich Nietzsche, *Twilight of the Idols*, 1889)

In 1997, the *British Medical Journal* published a paper with the title 'Sex and Death: Are They Related?'. It reported a study in the Welsh town of Caerphilly and some surrounding villages involving a group of nearly 1,000 men aged 45–59. The subjects were asked how often they had orgasms, and ten years later the data was analysed to determine whether orgasm rate had any effect on morbidity.

The mortality rate was found to be 50% lower in men with high orgasm frequency than in those with low orgasm frequency and the paper concluded that 'Sexual activity seems to have a protective effect on men's health.'

Despite the fact that the paper was published in the Christmas issue of the *BMJ*, which has a reputation for introducing some humour and relaxing its academic standards at that time of year, the idea that sex may be good for you proved very popular in the press, and articles and books have frequently appeared since then promoting similar messages.

In 1999, one author claimed that having sex frequently can make you look ten years younger, while in 2000 another report claimed with impressive-sounding accuracy that having sex at least twice a week can make your biological age 1.6 years younger than if you had sex only once a week. (Incidentally, the sixteenth-century founder of Protestantism, Martin Luther, advised that having sex more than twice a week was sinful, but less than twice a week was perhaps insufficiently respectful to our God-given procreative duties.)

What has been established is a correlation between frequency of sex and longevity, and between frequency of sex and looking young, but even if we accept that the correlation is generally true, the interpretation is not as clear as we might think.

Correlation

Does a high cholesterol level in the blood make a person more likely to have a heart attack? Are tall men more likely to marry beautiful women than short men? Does burning fossil fuels lead to an increase in atmospheric temperature? Does money lead to happiness?

To provide answers to such questions, statisticians draw up tables plotting one of the factors against the other. Then a fairly simple piece of mathematics, performed with one click on a computer, tells you not only whether the two are linked, but the extent to which this is so. The result of the calculation is a correlation coefficient between 1 and –1.

Two factors that are perfectly related to each other (i.e. they increase or decrease at exactly the same rate) have a correlation of 1; two factors that are totally independent of each other have a correlation of 0; a correlation of –1 indicates a perfect reverse relation (as one increases, the other decreases).

If we plot people's weight against their height, we get a correlation just over 0.4.

Shoe size and height correlate at a level of more than 0.9.

Studies correlating income and happiness have generally reached a conclusion that the level is around 0.4, but a study in the US in the early 2000s suggested that when annual income increases over $70,000, the happiness level changes very little, which suggests that it is not so much that money brings happiness as that poverty creates unhappiness.

Finally, I might mention a study in the *Journal of Sex Research* in 1993, which concluded that 'Height and foot size would not serve as practical estimators of penis length.'

When we detect a correlation between *A* and *B*, there are four possible explanations:

1. *A* (to some extent at least) causes *B*.
2. *B* (to some extent at least) causes *A*.
3. *A* and *B* are not directly causally connected, but there is a third factor *C* that may cause or be caused by both of them.
4. The whole thing has just occurred by chance, and *A* and *B* really have no connection at all.

When the amount of sexual activity is found to correlate with a longer life, perhaps we should not draw the conclusion that sex makes you live longer. Is it not perhaps more likely that people in good health have more sex than unhealthy people? Equally, instead of claiming that frequent sex keeps you looking young, it might be more reasonable to conclude that people who look younger are more successful at attracting sexual partners than those who look old and haggard?

Misattribution of cause and effect takes two forms: we can either get the two the wrong way round or we can believe there is a causal connection where none exists. In either case, this can lead to superstitions, irrational beliefs in lucky mascots and pseudoscience. The real trouble is that one of the greatest qualities of the human brain lies at the heart of such errors: our ability to spot patterns and form concepts. It's what happens *after* we have spotted a pattern – or think we have done so – that results in problems.

The great philosopher Karl Popper (1902–1994) argued that good science depends on a notion he called falsifiability. For a scientific theory to be worth taking seriously, it has to be formulated in a way that would allow experiments to disprove it if it is incorrect. Newton's Theory of Gravity made predictions about planetary motion that turned out to be correct: if observations had not supported the predictions, then the theory would have been shown to be wrong. Einstein's Theory of General Relativity, published in 1915, predicted that a large mass can bend light. He had to wait four years, but observations of a solar eclipse in 1919 confirmed his predictions: the mass of the Sun was found to have bent light from distant stars exactly as Einstein had calculated. More recently, the relatively new science of genetics and techniques for decoding DNA have supported Darwin's Theory of Evolution in a way that Darwin himself could never even have dreamt of.

A scientific theory can never be proved, however much evidence is gathered to support it, but as Popper pointed out, what makes a theory worthy of scientific interest is that it can be disproved if it is wrong. Most human beings, including most scientists, are not natural Poppereans. When we get an idea, we look for examples to support it, not examples to prove it wrong. Sometimes, we even go further and ignore counterexamples when they happen. As Erasmus Darwin wrote to his brother Charles in 1859 after reading *The Origin of Species*: 'The a priori reasoning is so entirely satisfactory to me that if the facts won't fit in, why so much the worse for the facts.' The evidence suggests that we have a strong tendency to share Erasmus Darwin's inclination to dismiss inconvenient facts that don't support our ideas.

According to Popper, when we have formed a hypothesis, we should collect data or conduct experiments that will tell us if it is wrong. If we are inclined to believe that walking under ladders is unlucky, the rational approach is to collect examples of both walking under and walking around ladders and see if any mishap has occurred shortly after. A simple calculation will then be enough to tell us how our ladder behaviour correlates with our fortune. In practice, however, people with a fear of walking under ladders will forget the many ladders walked under without bad consequences as well as all the mishaps that have occurred after walking round a ladder, and remember only those ladders that confirmed the superstition.

Equally, four-leafed clovers are remembered if they are followed by good luck, and three-leafed clovers are never credited with lucky accidents that may follow. As far as I know, no research has ever been done on the effects of finding four-leafed clovers while walking under ladders. I hypothesize that the good and bad luck would cancel each other out, but further experimentation is clearly needed.

A type of non-Popperean irrationality was demonstrated by the psychologist Peter Wason (1924–2003) in a delightfully simple experiment in 1960 illustrating a concept that later became known by the name 'confirmation bias'.

Wason gave a group of 29 psychology undergraduates, whom he described as 'intelligent young adults', the task of identifying a simple rule to which any group of three numbers might or might not conform. The subjects were given a set of three numbers that they were told did conform to the rule, and each of them was invited to suggest other groups of numbers that they thought might also conform to the rule, to see if they could discern it. Each time they suggested a group of three numbers, the experimenter would tell them whether these conformed to the rule or not. There was no time limit, but the participants were encouraged to find the rule with as few sets of numbers as they could. 'Remember that your aim is not simply to find numbers which conform to the rule, but to discover the rule itself,' they were told, and they were asked to write down the rule when they were highly confident what it was, '*and not before*'. The triple that they were given at the start was (2, 4, 6).

Most of the 29 supposedly intelligent young adults immediately suspected that the rule was 'three consecutive even numbers' but very few took proper Popperean steps to look for falsification of that belief. Instead, the majority of them sought evidence to confirm it. Some even restricted their efforts to establishing that triples such as (6, 8, 10) and (10, 12, 14) conformed to the rule, then leapt to the conclusion that their rule must be the right one without even bothering to test (3, 5, 7) to see if the rule applied also to odd numbers, or (2, 4, 28) to see if unequal distances between the numbers worked, or (6, 4, 2) to see if the numbers had to be increasing sequences or might be decreasing. There are plenty of potential rules that (2, 4, 6) conforms to, but the actual rule Wason had in mind was 'any three positive whole numbers in increasing order of magnitude'.

Only 6 of the 29 subjects settled on the correct answer at their first attempt. One gave up completely, which left 22 who came up with an incorrect answer; they were encouraged to continue trying. Ten of these got it right the second time, three more gave up, and nine gave another wrong answer. The next time, four of the nine were correct, three gave up and two were incorrect; neither of the two who were still wrong redeemed themselves at the fourth attempt and one more dropped out, leaving only one, who finally saw the light at the fifth attempt.

When even a group of psychology students show such a tendency to pursue false beliefs single-mindedly, it is hardly surprising that superstitions and false beliefs persist in the general population. Just look at all the supposed 'cures' for hiccups or the common cold that people swear by. It is often said that, when treated properly, a cold will last for at most seven days but if it is left untreated it may drag on for a week. If your cold clears up on day seven, it is natural human behaviour to believe that whatever you did on day six of those seven was the miracle ingredient that cured the cold. The next time you catch a cold, you therefore do it again, and again the cold clears up. Each time, your belief in the efficacy of your remedy, whatever it is, is increased. Karl Popper would surely have recommended trying something else, but most people would see that as too dangerous even to consider.

The problem is threefold:

1. We attach a greater importance to recent events than those in the more distant past.
2. We believe in samples of one so are reluctant to change a strategy that seems to work.
3. This leads to confirmation bias, which further strengthens our incorrect ideas.

When our mind, with its great talent for pattern recognition, spots an apparent connection, we impute causality to what may be only coincidence. The assessment of causality, however, ought not to be based on what our intuition tells us.

A particularly striking example of this occurred in the early 1990s when there was a growing amount of evidence that apparently suggested a causal connection between silicone breast implants and connective tissue disease: more and more women with the breast implants were found to be suffering from the disease. As the implants became suspected of being the cause of the problem, more newspaper articles appeared linking the two, and a large class action lawsuit was taken on behalf of the women sufferers against a leading producer of the implants. The implants were banned in the USA, a court found against the company and a massive settlement forced them to apply for bankruptcy. Some time later, sober medical research and proper statistical analysis concluded that the breast implants played no part in the development of the disease. It was true that many women with implants had developed connective tissue disease, but so had many women without breast implants, and many women with implants had no connective tissue problems.

Looking only at sufferers of the disease and discovering that many had implants is not enough to establish causality. One needs to know, at the very least, that the probability of the disease occurring in women after an implant is greater than the chance of its occurring with no prior implant, and the evidence did not support that conclusion.

Even when there is a genuine correlation between two things, it can be quite wrong to believe there is necessarily a causal connection. It has been shown, for example, that the number of Nobel Prizes won by a nation has a strong correlation with chocolate consumption. It does not follow that eating chocolate increases one's chance of winning a Nobel Prize.

In 2012, the *New England Journal of Medicine* published

a delightful paper by Franz H. Messerli MD with the title 'Chocolate Consumption, Cognitive Function and Nobel Laureates'. It included a graph plotting various countries' annual per capita chocolate consumption against the number of Nobel laureates that country had per million population. To give the graph added visual appeal, the points on the graph took the form of small national flags of the nations represented. Nearly all the flags lay very close to a straight line, from China (0.005 Nobel Prizes per million people and 0.08 kg of chocolate per person per year) to Switzerland (3.3 Nobel Prizes per million people and 10.2 kg of chocolate per person per year) showing a very strong correlation.

'The only possible outlier seems to be Sweden,' Messerli tells us. 'Given its per capita chocolate consumption of 6.4 kg per year, we would predict that Sweden should have produced a total of about 14 Nobel laureates, yet we observe 32. Considering that in this instance the observed number exceeds the expected number by a factor of more than 2, one cannot quite escape the notion that either the Nobel Committee in Stockholm has some inherent patriotic bias or that the Swedes are particularly sensitive to chocolate, and even minuscule amounts greatly enhance their cognition.'

He puts the result down to dietary flavonoids, which have been shown to improve cognitive function, and points out that a subclass of flavonoids called flavanols are widely present in cocoa.

Perhaps the most remarkable thing about the chocolate–Nobel correlation is that some people seemed to take it seriously and wrote alternative (but far less funny) interpretations of the result. The most complete and obvious of these pointed out not only that a nation's Nobels per million population correlated even better with the number of Ikea stores per million but that both Nobels and chocolate consumption correlated with per capita gross national product (GNP).

To show that they shared at least a little of Messerli's humour, the authors of that particular rebuttal of his findings did suggest that the chocolate–Ikea connection might possibly be taken as evidence that the supposed effect of chocolate on cognitive function might be what is needed to follow the Ikea instructions for assembling Ikea's flat-packs of furnishings, but they suggested this was unlikely.

Of course, it could be that whenever a country wins a Nobel Prize the winner, their friends and their university colleagues celebrate by eating more chocolate. In Messerli's flavanol theory, eating chocolate is the cause and Nobel Prizes are the effect, while the celebratory chocolate theory suggests it is the other way round.

Most likely of all, of course, is that a third factor, such as a nation's economic status, affects both Nobel Prize winning and chocolate consumption. Richer countries win more Nobel Prizes and eat more chocolate.

The phenomenon of confirmatory bias seems very human, but a very similar type of superstitious behaviour has been demonstrated in pigeons. The original experiment, later confirmed by many others, was conducted by the renowned behavioural psychologist B. F. Skinner in 1947.

Hungry pigeons were left in an enclosed area with a machine that dropped food pellets for them at regular intervals. Whatever a pigeon did between successive pellets became associated in its brain with the delivery of food and the pigeons were seen to develop complex behaviours, often involving hopping around in a specific direction a number of times, to ensure that the feeding continued. Shortly after it completed each dance, the food arrived, thus reinforcing the behaviour. By the end of the experiments, three-quarters of the pigeons had developed their own brand of superstition. And of course, the superstitions were justified by the arrival of the next meal.

Similarly irrational behaviour has been identified in human behaviour on entering a lift. Many lifts offer ideal scenarios for psychological experiments: you enter the lift, press a button for the floor you want and then wait for the doors to close and the lift to start moving. There is a delay, however, as you wait for the doors to close and another delay before the lift lurches into action. Many users develop forms of behaviour, not as elaborate or artistic as the pigeons' dances of course, that they believe make a beneficial contribution to the workings of the lift. Pushing the same button several times or holding a finger on it are often believed to cut down waiting time, while another common belief is that holding the choice-of-floor button down can make it more likely that the lift goes directly to the passenger's chosen destination without stopping at any intervening floor.

Just as with the pigeons waiting for food, the human lift passenger performs the action once, more out of impatience than anything else, and when the desired result happens, it becomes associated with that action, whatever it was. So the action is repeated next time and the belief in that association is reinforced every time.

On the same subject, there is a curious history about another button in lifts, which is the one that supposedly closes the door. These buttons came under great criticism when the Americans With Disabilities Act was passed in 1990 as the act specified that elevator doors had to remain fully open for at least three seconds to give disabled passengers time to enter the elevator. Door-close buttons were promptly disabled, and in the majority of US elevators they have no effect at all. Some psychologists, however, argue that such buttons function as a placebo to give the passenger an illusion of perceived control which diminishes stress. Passengers in UK lifts, however, are denied this placebo as British door-close buttons in lifts usually work.

Another example of illusory correlation that has a very long history and refuses to go away is the supposed effect of the full moon on human behaviour. The words 'lunatic', which dates back to the thirteenth century, and 'lunacy' (sixteenth century) derive from the Latin word for the moon, *luna*, and referred to a type of recurrent insanity that was thought to be dependent on phases of the moon.

As fans of horror films know, any human bitten by a werewolf is liable to change into a wolf when the moon is full. Long before such films, however, more literary examples abound of writers referring to the moon affecting human behaviour. In Shakespeare's *Othello*, it is the proximity of the moon rather than its fullness that the title character blames for his murder of Desdemona: 'It is the very error of the moon. She comes more nearer earth than she was wont. And makes men mad.'

In 1753, the judge and politician Sir William Blackstone, in his *Commentaries on the Laws of England*, drew a clear distinction between a lunatic and an idiot: 'A lunatic, or non compos mentis, is one who hath had understanding, but by disease, grief, or other accident, hath lost the use of his reason. A lunatic is indeed properly one that hath lucid intervals; sometimes enjoying his senses, and sometimes not, and that frequently depending upon the change of the moon.'

Such beliefs date back much further, with both Pliny the Elder and Aristotle suggesting that the moon's effect on the human brain may be analogous to its undoubted effect on the tides, as they both viewed the brain as our most moist organ. In modern scientific times, it has been pointed out that the gravitational effect of the moon on a person's brain is less than that of a mosquito perched on that person's arm, but such facts have not deterred some from investigating possible aspects of lunar lunacy.

Dozens of studies have been published supporting the view that certain aspects of behaviour, mostly undesirable, occur more frequently when the moon is full. In 1978, US psychologists David E. Campbell and John L. Beets published a paper entitled 'Lunacy and the Moon' which reviewed a large number of earlier papers relating lunar phase to types of behaviour, including psychiatric hospital admissions, suicides and homicides, and pointed out that few if any of their claims had been sustained in further investigations. They concluded that 'lunar phase is not related to human behaviour' and suggested that their findings 'should be sufficient to discourage future investigators from examining the lunar hypothesis'.

Unsurprisingly, future investigators were not discouraged. Only seven years later, psychologists James Rotton of Florida International University and Ivan Kelly of the University of Saskatchewan found it necessary to write an even more detailed follow-up report 'Much Ado About the Full Moon: A Meta-Analysis of Lunar-Lunacy Research' updating the earlier paper to add 19 more reports and three books, all supporting the Full Moon Theory (which detractors call the 'Transylvania Effect').

Their method was based on pooling the data in a total of 37 studies, many of which had involved small sample sizes, and applying statistical tests to the aggregated figures to see if there was a connection between moon phases and various types of lunacy, 'including mental hospital admissions, psychiatric disturbances, crisis calls, homicides, and other criminal offences'. Some statistically significant effects were indeed found but 'estimates indicated that phases of the moon accounted for no more than 1% of the variance in activities usually termed lunacy'.

Rotton and Kelly also mention the result of a survey they conducted in which '81 of 165 undergraduates (49%) indicated they believed that some people behave strangely when the moon is full'. They also mention a survey in 1973 of nurses working in a psychiatric setting which reported that 74% of them believed that the moon affects mental illness.

Their final words mention that it is common practice for such reviews to end by saying that further research is needed. 'We will break with this tradition,' they say, and express their hope that it has not only shown much ado but will also be 'much *adieu* about the full moon', expressing their forlorn hope that no more would be heard about lunar lunacy.

But werewolves are notoriously difficult to kill, and good science continues to struggle against a belief in the potential evils wrought by the full moon. One of the latest pieces of research, however, claims a surprisingly different effect.

In 2018, the *British Medical Journal* published a paper 'Lunar Cycle in Homicides: A Population-based Time Series Study in Finland'. Analysing 6,808 homicides in Finland between 1961 and 2014 for which precise dates were available, Simo Näyha, Professor of Health Science at the University of Oulu, Finland, counted the numbers of homicides that occurred during each of the eight moon phases. Previous studies had suggested that more crimes in general occur during the full moon, but Näyha's startling result was that Finland experienced fewer homicides during full moons than any other moon phase.

'Why homicides decreased during the full moon is not easily explained,' he admits:

> One might speculate that the full moon ... may have some superstition-based meaning in peoples' minds that refrains potential lunar phobic perpetrators from committing the act ... The victim's behaviour might play a role, too ... Potential homicide victims who feel themselves threatened may avoid moonlight to protect themselves, or they may believe that something unfortunate could happen during the full moon.

He also suggests that a drop in homicides could result from 'an atavistic remnant from the animal kingdom where certain prey animals suppress their activity in moonlight to hide themselves from predators, or perhaps, their enhanced visual acuity in moonlight would help them detect predators'.

One might also suggest that as previous studies had allegedly shown an increase in crime when the moon is full (possibly because the increase in moonlight gave greater opportunities for crime), it could be that potential murderers were too busy committing other crimes to commit homicides, or indeed that they had turned into werewolves and were too busy howling at the moon. Further research on these matters is, I would strongly suggest, not needed.

The belief in lunar effects on lunacy looks strongly like another case of confirmation bias which experience suggests may affect not only superstitious people but scientists and other academics who ought to know better. In 2005, the Greek–American physician John Ioannidis published a delightfully controversial paper in the *PLOS* (Public Library of Science) *Medicine* journal under the provocative title 'Why Most Published Research Findings Are False'. His argument, supported by some statistical calculations, was that a number of elements combine to produce the effect that he spells out in the title.

First and perhaps most significant are aspects of bias influencing the direction of research. Scientific experiments tend to begin with a hypothesis, and scientists believe their intuition in that respect tends to be correct. So they do not question the methodology if the result supports their predictions, but repeat the experiment, often under slightly altered conditions, if it does not do so. Furthermore, the editors of scientific journals are far more likely to publish papers confirming expected results rather than those suggesting they may be incorrect. Both the researchers' designs and the editors' choices show aspects of confirmation bias.

Add to that the fact that several research groups may be doing essentially the same experiments, in many cases the sample sizes may be too small for the results to be absolutely convincing, and a large number of statistical tests may be performed of which only the successful ones, which could have occurred by chance, will be submitted for publication. So it should be no surprise that many published results fail to be replicated in later experiments. As Ioannidis puts it: 'for many current scientific fields, claimed research findings may often be simply accurate measures of the prevailing bias'.

Naturally, Ioannidis's paper was greeted by howls of indignation from several researchers who suggested that much of what he was saying to support his thesis was itself suffering from confirmation bias, but even the most critical of them agreed that there could well be something in what he claims.

On the general topic of spurious correlations, we should mention that if you perform enough tests, some (in fact around 1 in 20, taking the usual $p < 0.05$ criterion) will turn out to be statistically significant. Harvard law student Tyler Vigen has written a computer program that looks for correlations in all sorts of totally unrelated data and has collected some of the most startling results in his book *Spurious Correlations*. Here are a few of my favourite examples with my own tentative conclusions:

The annual per capita consumption of cheese in the US from 2000 to 2009 had very high correlation with the number of people who died by becoming tangled in their bedsheets.

Conclusion: eating cheese causes convulsively bad dreams.

The divorce rate in Maine from 2000 to 2009 had almost perfect correlation with the US per capita consumption of margarine.

Conclusion: Eating butter keeps a married couple together.

The number of honey-producing bee colonies in the USA correlated negatively from 1990 to 2009 with the number of juvenile arrests for marijuana possession.

Conclusion: Feeding children with honey keeps them off drugs.

CHAPTER 9

Percentages and More Misleading Mathematics

More natural mistakes

Removes up to 100% of dirt, grease and grime

(claim on bottle of cleaning liquid)

As advertisers and politicians know, nothing can mislead as subtly as a percentage. Of course a cleaning product removes 'up to 100% of dirt'. It can hardly remove more than 100%, and anything else satisfies the 'up to 100%' criterion.

If you want a small number to seem large, express it as a percentage. Saying that 13 UK prime ministers were educated at Christ Church College, Oxford, somehow sounds less impressive than reporting that 23% of prime ministers since Walpole studied at Christ Church.

If you want a small percentage to seem large, express it as a number. Hearing of plans to recruit 5,000 nurses every year sounds better than plans to increase the number of nurses by 1.6%.

On the other hand, small numbers of people make a greater personal impact than percentages. Knowing that 1,784 people were killed on Britain's roads in 2018 makes less of an impression than being told that one person dies in a traffic accident every 4.9 hours. That's less than the time between lunch and dinner, and that one person could be me. As Josef Stalin famously said: 'One death is a tragedy; one million is a statistic.'

Percentages are also a good way to hide a small sample size. When a cosmetics company reported in a TV advert that 71% of women found that their product visibly reduced cellulite, viewers may have been impressed. The small print at the bottom of the screen, however, revealed that those 71% were just 34 women from a sample of 48. Would the viewers have been so impressed if they had seen the small size of the sample?

I must admit that I was a little puzzled by a similar advert referring to 'clinical tests' of a cream showing that 93% of women felt it made their skin softer, and 79% said their skin felt firmer. The sample was supposedly 40 women, which means the 93% is clearly a rounding up of the 92.5% figure referring to 37 of them, but the 79% is not easy to explain. When you do the sums, you see that 32 people would have been exactly 80% while 31 is 77.5% which rounds up to 78 not 79.

I often perform similar calculations when I see sample

sizes, to check whether the percentages make sense, and it is rewarding to find instances when they do not. The Advertising Standards Agency (ASA) does not rule on such matters; it suggests that sample sizes need not be included in marketing, unless it would be misleading not to do so. Their guidelines say: *'if results are based on a robust sample size in which statistically significant findings can be drawn, then there is no need to include the sample size in the ad'*. The facts that the advertisers in both the cases above did include the sample size suggests that they had doubts about the robustness of their sample sizes. I certainly agree with them. These sound like very small samples and we do not know how they were selected.

Misleading uses of percentages in advertisements go deeper than this, however. What should we make of a toothpaste that is apparently recommended by more than 80% of dentists? If you are unimpressed by this claim, then you are not alone: the ASA ruled that the advert was in breach of the Committees of Advertising Practice policies on Substantiation, Truthfulness, Testimonials and Endorsement of medicines and demanded that the makers withdraw the advert. The problem was not only that the dentists in question were not told the survey was for an advertising campaign or that their recommendations might be used as an endorsement, or even that the question they were asked encouraged them to name as many toothpastes as they wanted in their recommendations. The 'Truthfulness' breach came from the ASA's considered view that most people would think that the wording on the advert meant that 80% of dentists recommended that particular toothpaste *above all others.*

This is an illustration of what is perhaps the biggest problem of all in interpreting the numbers in adverts: when they are based on responses to surveys or other questionnaires, we are never told precisely what questions were asked.

What should I make of it when a survey reports, as I read in 2019, that 46% of people choose porridge as their top breakfast option for winter? Personally I like nothing better on a cold winter's morning than a bowl of porridge cooked in milk with sugar and blueberries and then drenched with a quick sauce made by warming whisky and honey, but I doubt that almost half the nation agree with me. In fact, I can hardly think of anyone I know who would say that porridge is their top breakfast option even in winter.

I strongly suspect that the people surveyed were asked to list all the breakfasts they enjoyed, and 46% of them included porridge in the list. But even if porridge was mentioned in more such lists than any other breakfast food, that does not make it the top breakfast food for 46% of the nation. And while we are on the subject, what exactly is meant by '46% of people'? Even if we assume it refers to people in the UK, we should ask whether it is 46% of all people or 46% of adults, or 46% of adults who have breakfast. Around 18% of the population are under 15, and children are not widely noted for their love of porridge.

When we see any reference to a certain percentage of people, we should ask who those people are. Does it refer to the whole country or is it restricted to those with a specific characteristic? If it is a survey, who were the sample and what was done to ensure that they were representative of the target group? Were they volunteers and, if so, do we have any reason to believe they are representative of the average? Were the don't-knows and didn't-replies excluded *before* the percentages were calculated? It can make a great deal of difference on matters where many people did not answer.

Finally, we must ask whether we believe the responses people give to questionnaires or pollsters. In 2017, a survey reported that men are 30% more likely than women to have had a sexual encounter at the office Christmas party. The obvious conclusion to jump to, which the report clearly suggested, was that male office workers are more lascivious than women, but that's not what the figures say at all.

Just think about it: if we are restricting our attention to heterosexual relations, then every sexual encounter involving a man also involves a woman. So if there are equal numbers of men and women attending the parties, and the men are 30% more likely than the women to have had sex, then the average woman must have had sex with at least 1.3 men (more if some of the men are also not monogamous).

On the other hand, if everyone partners at most one person they encounter at the party, then for the men to be 30% more likely than the women to have an encounter, there must be 1.3 times as many women as men at the party in total.

Besides these explanations, there are two other possibilities: some of the men form gay relationships at the party, or men tend to exaggerate reports of their sexual success.

Such male exaggeration, if that is the case, is not restricted to reports of Christmas parties. In the 1990s, a survey in Quebec reported that men had an average of 10.8 sexual partners in their lives, while women had 6.2. In France, the figures were 10.1 for men and 4.4 for women, while the USA reported 11.5 and 5.0 and the UK had 12.7 lifetime partners for men and 6.5 for women.

If these figures are true, then in all these places the average man must find half his lifetime sexual partners in countries other than those surveyed. Exaggeration by men or reticence in giving the true figures by women seems by far the more likely explanation.

In 2003, a study in the USA revealed what had been happening. A sample of over 200 heterosexual students were divided into three groups and all were asked to fill in a questionnaire asking about their number of sexual partners. The first group had their names at the top of the questionnaires and were told the questionnaires might be read by the researchers; the second group replied anonymously; and the third group had electrodes placed on their hand, forearms and neck and were told they were attached to a polygraph lie detector.

In the first group, the men reported significantly more sexual partners than the women. That was also true, though to a lesser extent, for the second group. In the third group, however, who thought their responses were being tested by a lie detector, the male–female difference disappeared. The surprising thing was that the females were found to have been underestimating their replies considerably more than the men had been exaggerating theirs.

Here are a couple of percentage problems to try. If you get them both right, you will detect at least half the errors involving percentages that you are likely to encounter.

An item increases its price from £1 to £4. What percentage increase is this?

(a) 300%   (b) 400%

What percentage decrease would then be necessary to take the price back down from £4 to £1?

(a) 300%   (b) 400%   (c) 75%

The correct answer to the first question is (a) 300%, but it is remarkable how often newspaper reports get such things wrong, particularly when they report a doubling in price as 'a 200% increase'. A 50% increase increases something by half its former value; a 100% increase doubles the price; a 200% increase triples it and so on.

The second question produces an even more typical error. What many fail to grasp is the idea that a 100% increase is reversed by a 50% decrease. This seems totally unfair, but the logic is unassailable. If you double something, that's a 100% increase, but to reverse it then involves halving, which is a 50% decrease. The correct answer to the second question is (c) 75%. The price is reduced by £3, which is 75% of £4.

Beware of any company that tells you that its annual profit has risen 20% on last year's figures, which cancels out the 20% fall the year before. A 20% fall would have reduced £100 to £80. If that figure then increases by 20%, we are left with £96, which is still £4 less than the original amount. To restore a 20% fall, one would need a 25% rise.

Now look at the following extract from a recent news report: 'The total number of unauthorized migrants crossing the US–Mexico border has dropped by 300% over the past 16 years.' Sounds impressive, doesn't it? But actually it's crazy. A 100% drop would mean that migration had totally stopped. The only possible interpretation of a 300% drop would be that twice as many people had migrated from the USA to Mexico as the previous total had been for migration in the opposite direction.

Here's another one, from the National Democratic Alliance in India: 'Communal riots decreased by 200% under NDA rule.' Again a 100% reduction would have reduced rioting to zero. To talk of a 200% decrease would involve the concept of a negative riot and I cannot even speculate what that might be.

Both these examples have at their heart the mistaken idea that an increase of a certain percentage is balanced by a decrease of the same number. As our earlier question made clear, that's not right. They may be the same percentages, but they are percentages of different numbers.

While we are on the subject of misleading percentages, what on earth are we meant to make of a disinfectant that proudly claims to kill 99.9% of all bacteria? Is that 99.9% of each individual type of bacteria or 99.9% of all different types of bacteria?

Bacteria are microscopic, single-celled organisms that exist everywhere. Some are potentially harmful to humans but the vast majority are either harmless or beneficial to both plant and animal life. The good bacteria inside our bodies are even helpful in restricting the effects of bad bacteria. We hear so much about bacterial infections that bacteria in general have a bad name, but something that killed off 99.9% of all bacteria, as the disinfectant advert promises us, could be disastrous.

Perhaps this is a good moment to introduce one of my favourite numero-linguistic aversions. If A is five times bigger than B, is B five times smaller than A?

I have often heard phrases such as 'five times smaller' and I know what it is intended to mean, but I feel there is something inherently illogical about it. The word 'times' indicates multiplication. If A is five times the size of B, then to work out the size of A we multiply the size of B by 5. Simple. But if B is smaller than A, there is not a measure of smallness that we can multiply by 5. We need to divide by 5 or multiply by one-fifth: B is one-fifth the size of A.

The problem stems from the expression 'five times bigger than' which I find puzzling enough on its own. We used to say 'five times as big as', but I think people got confused by 'as big as', which on its own implies equality of size, so people started saying 'five times bigger than' to emphasize the 'bigger than' aspect. The trouble is that '100% bigger than' means 'twice the size', so '200% bigger than' ought to mean three times the size. But 200% is equal to twice, so by that logic 'five times bigger than' strictly ought to mean 'six times as big'.

I shall spare you the numero-linguistic problems introduced by the question of whether Blackpool Tower is twice as small as the Eiffel Tower. Let us return to the questions one should ask when reports include numerical comparisons.

The first question concerns possible selectivity in the choice of data being compared. Some years ago, the UK rail network proudly announced that trains had achieved their best punctuality figures for five years. That immediately made me suspicious. If they say 'best for five years', I naturally began to wonder about what happened six years ago. Finding data that went back ten years, I discovered that a disastrous rail accident five years previously had resulted in speed restrictions, track repairs and signalling improvements, all of which resulted in slowing trains. This had severe effects on punctuality, which had gradually improved but was still not back at the level it was six years ago. The latest punctuality figures were indeed the best for five years, but were still not as good as six years ago.

There is also the question of whether we are making a fair comparison: had the definition of 'punctual' changed in the period under consideration? In 2013, the Swedish Transport Administration claimed improved punctuality rates for Swedish trains – but the time for what counts as a delay had been tripled from 5 to 15 minutes.

In the UK also, changes have been made in the definition of punctuality. What if a train arrives at its final destination close to the scheduled time but was late at all the stops on the way? What if operating difficulties such as a missing crew member caused the train to be rescheduled but it then arrived at the correct time according to the new schedule? Both these little nuances have been used in the past to improve punctuality figures, but the delayed passengers tend not to be impressed.

Just in case you have not had enough of percentages, let me round off this chapter with some examples from surveys I have seen in 2020. There have been more than enough such surveys for me to pick one example of every percentage from 1 to 100, which should leave you needing no more percentages, at least until next year. Unless otherwise specified, all the surveys were conducted in the UK.

1% of people say their children drink raw milk.

2% of people say they never compliment anyone.

3% of our total household expenditure is on cafe and restaurant meals.

4% of the time of wedding photographers is spent taking photographs.

5% of dogs in Finland suffer from separation anxiety.

6% of people worldwide eat seafood every day.

7% of people say their level of educational attainment prevents them from getting ahead.

8% of 18 to 24-year-olds buy more than ten items a month from online fast fashion retailers.

9% of people think blue is the best colour for living rooms and hallways.

10% of children leave primary school unable to brush their teeth.

11% of people regularly sleep soundly through the night.

12% of people regularly sell their second-hand clothes.

13% of people are unwilling to share a packet of crisps.

14% of students struggle to get their accommodation deposits back.

15% of office workers do not like spending money on charity requests from colleagues.

16% of adults had an imaginary friend when they were young.

17% of farmers have access to superfast broadband.

18% of people think grey is the best colour for living rooms and hallways.

19% of American women have gone shopping when drunk.

20% of shoppers regularly miss deliveries.

21% of adults are afraid of needles.

22% of parents hide fruit and veg in their children's food at least once a day.

23% of teenagers say they are keen to pursue a career in the green economy.

24% of people are unhappy at work.

25% of the world population is under 15.

26% of people say gambling livens up life.

27% of men do not read books.

28% of employers find visible piercings in a potential employee distracting.

29% of people say it would be better if gambling were banned.

30% of farmers have download speeds of 2 Mbps or less.

31% of people are worried about food hygiene when eating out.

32% of people think the British Empire is something to be proud of.

33% of students renting accommodation are troubled by damp.

34% of children voluntarily pick up someone else's litter.

35% of adults in employment suffer from relationship stress.

36% of people worldwide have slept separately from their bed partner in order to improve sleep.

37% of women say they lie at least once per day.

38% of 8 to 17-year-olds find the Internet a safe place to be.

39% of workers are not satisfied with their salaries.

40% of adults consider themselves coffee drinkers.

41% of Britons would put money aside to be able to travel to space imminently.

42% of adults work more than 30 hours a week.

43% of landlords are unaware of the Homes (Fitness for Human Habitation) Act 2018.

44% of people are worried about food prices.

45% of people have concern about food safety in restaurants.

46% of Europeans think air pollution is an important environmental issue.

47% of adults consider themselves tea drinkers.

48% of women say they would be confident to tell off a senior colleague for making a sexist comment.

49% of Americans have more savings than credit card debt.

50% of men say they lie at least once per day.

51% of managers have knowingly discriminated against a potential employee because of the way they looked.

52% of men say it's acceptable to ask a work colleague for a date.

53% want their local authority to be able to charge a small fee to help support tourism.

54% of office workers say they are happy to put money in for a colleague's leaving card or present.

55% of consumers are concerned about sugar in food.

56% of the world population live in cities or urban areas.

57% of car drivers say the issues around running an electric car are too daunting to make them buy one.

58% of teachers feel they are underpaid compared with other graduate professionals.

59% of the world population actively use the Internet.

60% of children aged three to seven multitask while watching TV.

61% of Britons when eating out like to order something they have had before.

62% of Britons think we should still have a monarch.

63% of people think 'Best Before' and 'Use By' labels should be scrapped.

64% of people lose sleep over work worries.

65% of 25 to 49-year-olds speak to their voice-enabled devices at least once a day.

66% of people are looking for a better work/life balance.

67% of football fans think matches are less enjoyable since the introduction of VAR.

68% of people have donated to charity in the last three months.

69% of Americans think it is unlikely that aliens will visit Earth this year.

70% of consumers say ginger has a positive effect on their health.

71% of Arizona voters think climate change is a serious problem.

72% of people think 'Best Before' and 'Use By' labels are too cautious.

73% of people say they would be prepared to pay more tax to boost spending on the NHS.

74% of women have never travelled solo to a continent other than Europe.

75% of Leicestershire Police officers think all colleagues should be issued with a taser.

76% of people feel happy or very happy.

77% of people do not get their left and right muddled.

78% of adults say they have never had a bad Valentine's Day experience.

79% of people have never worn their pyjamas in public.

80% of the world's known species are insects.

81% of people are worried about meat quality standards in future trade deals.

82% of people want extinct species reintroduced to the UK.

83% of working people aged 65 or over say they are satisfied with their jobs.

84% of people say working hard is the key to 'getting ahead in life'.

85% of women are wearing the wrong bra size.

86% of adults would not be put off ordering their chosen dish in a restaurant if their companion ordered it first.

87% of Britons eat meat.

88% of people agreed or strongly agreed that they feel safe in West Sussex.

89% of Mexicans do not have confidence in Donald Trump to do the right thing in world affairs.

90% of people worldwide are biased against women.

91% of adults never use a tanning bed.

92% of people have donated to charity at some time in their lives.

93% of the Chinese choose public transport over cars for environmental reasons.

94% of consumers report a better product offering online than in-store.

95% of EU citizens are worried about the impact of chemicals in products on the environment.

96% of Americans earning over $100,000 a year are satisfied with their personal lives.

97% of full-time illustrators are proud to be in the industry.

98% of parents are concerned about their children's privacy online.

99% of people check their email every day.

100% of all percentages ought to be viewed with at least a modicum of scepticism.

Before giving any credence to the figure, ask yourself: what is it a percentage of; how did the researchers obtain their sample; how many were in the sample; was it likely to be typical of the general population; did they exclude 'don't knows', 'not applicables' and other non-respondents before calculating the percentage; what other questions were asked; was the survey performed by a team or company that had a vested interest in the findings or was it truly independent; and is it likely that the respondents, for whatever reason, did not answer honestly? Only when you have the answers to all those questions will you be in a position to judge what the figures really mean.

CHAPTER 10

Chaotic Butterflies

The mathematics of chaos, catastrophe and complexity

The ability to be totally confused at times is important, and for everything to fall into chaos and really not know what to do. Chaos can be incredibly creative.

(Mark Rylance, *Daily Telegraph* interview, 31 December 2016)

In 1985, the Royal Society, the Royal Institution and the British Association for the Advancement of Science jointly expressed concern about the level of scientific understanding among non-scientists by founding the Committee on the Public Understanding of Science (Copus). Their aim was to make advances in science more accessible to people.

This objective quickly received government funding from the Office of Science and Technology and, in their own good time, professorships in the Public Understanding of

Science were endowed at a number of universities. Richard Dawkins was installed as the first Simonyi Professor for the Public Understanding of Science at Oxford University in 1995, a post he held until he was succeeded by Marcus du Sautoy in 2008.

Meanwhile, doubts had been growing about the expression 'public understanding', as reported by the government Select Committee on Science and Technology in 2000: 'We have been told from several quarters that the expression "public understanding of science" may not be the most appropriate label.' It quoted the Chief Scientific Adviser to the government, Sir Robert May, as calling it a 'rather backward-looking vision'.

The report argued that the words 'public understanding' suggest 'a condescending assumption that any difficulties in the relationship between science and society are due entirely to ignorance and misunderstanding on the part of the public'. It went on to say that: 'It is also increasingly important that scientists should seek to understand the impact of their work and its possible applications on society and public opinion.'

In view of such objections, I suppose the Committee could have set up a new Committee for the Public Understanding of Public Understanding, but instead they suggested a change of name and composition of the existing committee; Copus was discontinued in 2002.

The real problem with the public understanding of science seems to me to be that science has become increasingly difficult for the public to understand. The ancient Greeks may have been happy enough to accept that solid chunks of matter were all composed of atoms (meaning a particle incapable of further division), but when Ernest Rutherford split the atom in 1917 it stretched public understanding, and when quantum theory began to be developed in the mid-1920s, the last vestiges of true public understanding quickly disappeared.

How could the public be expected to accept or understand that particles can be in two places simultaneously, or that light is sometimes a wave and sometimes a particle, or that a cat can be both alive and dead at the same time, as Schrödinger told us?

If any university decides to inaugurate a Professorship for the Public Misunderstanding of Science, or a Professorship for the Gross Oversimplification of Science, I should be delighted to apply. These, I think, are realistic ambitions. Public misunderstanding, however, is a problem not just for science, but also for mathematics, and there can scarcely be a better example than the way Chaos Theory and the so-called 'Butterfly Effect' have been represented to mass audiences.

Chaos Theory is a recently developed branch of mathematics that has induced a subtle change in the way we view the world, but you would hardly think so from the way it has generally been presented in the popular media, particularly in films.

In 2004, the film *The Butterfly Effect* began with the following words appearing on the screen: 'It has been said

Chaos Theory

Chaos Theory is the science of predicting the unpredictable. Or, to be more precise, detecting when the conditions of the weather, the stock market, the turbulent flow of a fluid, or many other phenomena make accurate prediction impossible.

In such circumstances, the next best thing is being able to recognize when this happens. In other words, being able to accurately predict that accurate prediction is impossible. Particularly in the realms of stock market investment and weather forecasting, the prediction of chaos has become vital.

The Butterfly Effect

The important thing to remember is that there is a big difference between the genuine Butterfly Effect and the popular misconception about it. What they have in common is the idea that the flap of a butterfly's wing in one part of the world can make the difference between good weather and a typhoon in another. The difference is that the popular misconception is to see this as a warning that a tiny feature may grow to enormous proportions; the far more interesting concept is that, in certain circumstances, initial conditions may be so sensitive to small changes that no matter how accurately we measure them, the

> calculated results may differ hugely. If a butterfly's wing flap can result in a typhoon, another butterfly's mere exhalation may stop it.

that something as small as the flutter of a butterfly's wing can ultimately cause a typhoon halfway round the world' and, beneath that statement, the words 'Chaos Theory' appeared.

In 2005, another film appeared called simply *Chaos* in which a highly devious bank robber played by Wesley Snipes introduces the theme by telling a detective: 'Chaos has some order to it.'

Another detective tells his colleague that he believes that Chaos Theory may hold the key to the criminal's intentions and he explains that the theory is: 'the study of phenomena that appear random but in fact have an element of regularity that can be described mathematically. The initial state of events may seem unregulated and random but eventually a pattern emerges and in the end all the pieces fit together.'

Even in the much earlier film *Jurassic Park* (1993), Jeff Goldblum's character cheerfully attributes a wide range of unpredictable outcomes to the Butterfly Effect and uses 'chaos' as a near synonym for unpredictability. These examples, and many more, turn the mathematics of Chaos Theory on its head in misguided attempts to simplify it, as the real history of its development shows.

The ancient Greeks gave the name 'Chaos' to the vast void out of which the Earth was created. According to the pre-Socratic philosopher Hesiod, who lived around 700 BC, Chaos was the first thing to exist but could also be seen as the gaping void created when the Earth and Sky were separated from the Heavens.

The Romans had a similar view. Ovid's *Metamorphoses*, written in the first century BC, described Chaos as an unformed mass, a shapeless heap formed of a jumble of all the elements. We may now talk of countries, political systems or protest movements 'descending into chaos' but, to the ancients, Chaos was the primeval mush out of which order was created. Appropriately, the question of whether Chaos is seen as a starting point or a conclusion also lies at the heart of the confusion about the theory that takes its name.

The foundations of Chaos Theory were laid by the American meteorologist Edward Lorenz in 1961. Having started his professional life as a mathematician, he had naturally enough acquired an early computer and wanted to use it to help predict the weather. Newton's Laws of Motion had long ago taught us that the future behaviour of any physical system can be predicted if we know the masses, locations and velocities of all its components at a given moment; and the atmosphere is, after all, one admittedly huge and complex such system.

Lorenz therefore produced a mathematical model of the weather, taking only the variables that he considered most important, gave them all plausible initial values and set the computer to work out what was going to happen in the future. The results were encouraging: they looked like real weather patterns, but then he had a lucky, and highly influential, accident.

Having had one good run of figures, he decided to run it again but for a longer time period. To save time, however, he started the new run halfway through the old, entering figures for the new starting point from the computer printout of the old one. That printout, however, gave figures to three decimal places only, though the computer's memory held six.

For a short time, the new run gave much the same figures as the old, but then they began to diverge significantly, and the divergence grew and grew. After checking his figures, Lorenz came to the only possible conclusion: for nearly 300 years, we had been missing something vital in Newton's Laws of Motion. We knew that accurate measurement of masses, locations and velocities was essential to predict future behaviour, but we had assumed that the more accurate our measurements were, the more accurate the predictions would be. Lorenz's simple system had shown that even when figures differed by only one part in 10,000 (which was the difference between the three decimal places in his printout and the six in his computer), the differences in the resulting predictions could soon grow enormous.

Lorenz's first reaction was to write a paper for the *Journal of the Atmospheric Sciences* with the unprepossessing title 'Deterministic Nonperiodic Flow', which cast doubt on the feasibility of long-term weather forecasts. Thanks to Newton and later laws of fluid motion, the system was governed by precise mathematical rules that made its behaviour predictable for all time if we knew the initial conditions. But that last condition was a real stumbling block. If we measured the initial conditions to three decimal places, the results showed that six could have given a different answer. And if we had been accurate to six decimal places, the seventh might have altered everything.

This demonstration of the potential unpredictability of deterministic systems was the idea at the heart of Chaos Theory and it changed everything. It wasn't, as Wesley Snipes said, that chaos has some order to it but more that order may result in chaos. And it was not that phenomena that appear random have an element of mathematical regularity, but that predicting phenomena we know to be regular may veer off into apparent randomness.

No matter that Wesley Snipes called himself by the name 'Lorenz', and that a plane ticket was later booked in the name of James Gleick, who was the author of the book that popularized Chaos Theory, the chaos of the film was far from the chaos unearthed by Edward Lorenz.

All of which leaves the question of how butterflies muscled in on the action and took such a prominent part in the public misunderstanding. The answer to that question lies in a blend of three unrelated factors.

The first was a diagram produced by Lorenz to illustrate the chaotic path a particle might follow when under the influence of several forces. Traditional assumptions were that the path might vary greatly in its early stages but would settle down to something regular and predictable. Instead, his mathematical model plotted its path as consisting of two spiral formations with the particle swinging round and round one of them before shooting off onto the other one and eventually back again, this procedure going on for ever. The two spirals, angled away from each other, were rather reminiscent of a butterfly's wings.

At that stage, the term 'Butterfly Effect' had not yet taken root, but the two spirals may have contributed to that name when it was first associated with Chaos Theory in 1972. To be precise, it was on 29 December 1972 when Lorenz presented a talk at the 139th meeting of the American Association for the Advancement of Science held in Washington, DC, entitled 'Predictability: Does the Flap of a Butterfly's Wings in Brazil Set Off a Tornado in Texas?'.

Lorenz had in fact submitted his lecture without a title, and that imaginative line about the butterfly was suggested by fellow meteorologist Philip Merrilees. Curiously, the butterfly had begun life as a seagull. As Lorenz himself wrote in 1963 when describing his early findings: 'One meteorologist remarked that if the theory were correct, one flap of a seagull's wings would be enough to alter the course of the weather forever.' Changing the seagull into a butterfly was a masterstroke, beautifully exemplifying the possible effect of one factor so tiny that it cannot possibly be taken into account in practical terms.

Even if the wing flaps of all the world's butterflies could be incorporated into a computerized model of the atmosphere, one of the butterflies farting might lead to a totally different prediction. I am not sure whether butterflies do fart. Further research is clearly needed on that topic, but it does not matter: the merest exhalation of breath by the butterfly could be enough to throw the results into chaos. As Lorenz pointed out, however, if a butterfly's wing flap in Brazil could result in a tornado in Texas, it could just as easily prevent such a tornado.

To add to Lorenz's butterfly wing spirals and that tornado-inducing (or preventing) butterfly wing flap, a third reference completes the trinity of chaotic butterflies and this one predates Lorenz's discoveries by a decade. In 1952, the American science fiction writer Ray Bradbury wrote a short story called 'A Sound of Thunder' set in a future age of time travel. The tale featured a wealthy adventurer named Eckels who takes a trip back to the Late Cretaceous period in order to go on a hunt to kill a Tyrannosaurus rex. The organizers of the hunt have carefully arranged everything to ensure that nothing happens that might change history, and the hunters are told firmly to stick to a pre-determined path in order to kill a dinosaur that would have died within minutes anyway.

On seeing the tyrannosaur, however, Eckels panics and flees off the path. The organizers are furious and try to undo any damage he might have caused, but Eckels later discovers a crushed butterfly on the sole of his boots. When they return to their own time, they find that everything has subtly altered: the English language, people's behaviour and, most of all, politics have changed for the worse. And all because of that butterfly.

We do not know whether Lorenz was even vaguely aware of the Bradbury story, but it fitted well with his thesis that a butterfly can change everything. Lorenz's work may have been restricted to meteorology, but the revelations of his Chaos Theory quickly spread to other disciplines for which there had previously been no explanation of apparently random behaviour. The mathematics of chaos explained the phenomenon of turbulence in fluid mechanics. Geometric chaos led the mathematician Benoit Mandelbrot to develop wonderfully intricate designs that he called 'fractals', endlessly repetitive but on ever smaller scales, which show how simple finite curves can be insufficient to describe intricate structures such as coastlines. In fact, by increasing the detail of the detailed squiggliness of a coastline, you can make it as long as you want. The more squiggles and the finer they are, the greater the measured length, without limit.

Chaos Theory was employed to investigate and explain unexpected lurches in population dynamics, stock market fluctuations and other economic predictions. It even appeared in the design of some executive desktop toys, including a simple but revealing gadget consisting of a metal ball X on the end of a pendulum suspended above a tray holding three magnets A, B and C (see diagram).

How to draw a curve of infinite length

1. Start with a straight line of which the middle third is replaced by two sides of an equilateral triangle. If the original length of the line was 3*s*, the total length now is 4*s*.

2. You now have four straight pieces of equal length. As before, replace the middle section of each by two sides of an equilateral triangle. There are now 16 line segments, each of length $4s/3$.

3. Do the same thing again.

4. And again ... and again.
Every time the total length is multiplied by $4/3$.

5. This is why coastlines can become as long as you want if your measurements take into account smaller and smaller squiggles.

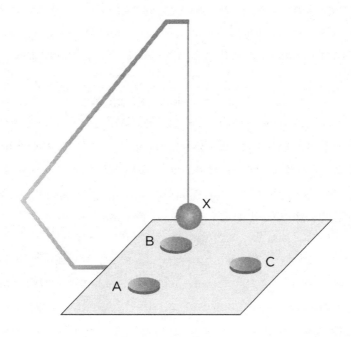

You pull the metal ball (X) back, let it go and watch it meandering, pulled to and fro by the magnets (A, B, C). It may start apparently happy to swing between two of the magnets, then the third catches it and sets it off on a new path. Eventually, as it loses momentum, the ball is captured by one of the magnets and its journey ends.

In fact computer models of this game can track the path the ball follows for any initial position. Since each such position is directly above a unique position on the plane of the magnets, that plane can be shaded in three colours corresponding to the three magnets. The results show that, for some clearly defined regions, the ball will end up at magnet A, while other regions correspond to magnets B and C. But at certain points the three colours get closer and closer together until they can no longer be differentiated.

In practice, what this means is that you can start the pendulum off somewhere, let it swing and see where it lands, but when you try to start the ball at the same spot, you cannot be sure of replicating the path no matter how close you get. In the regions around such points, the ball's behaviour becomes chaotic. No matter how close to the original point you come, it may lead to a different result.

This has led to a significant change in perception of the basic problem. For centuries, we had seen the difficulty as one of making sufficiently accurate measurements. Now we realized that, however accurate our measurements might be, they may not be accurate enough for even an approximate prediction to be made. Attention therefore shifted to the question of identifying whether we were on the verge of chaos or, to put it more scientifically, whether our measurements are being taken in a region that is arbitrarily sensitive to initial conditions.

In the past, predictive computer modelling consisted of writing a program to simulate the relevant laws of physics, plug in all the available data, run the program and see what happens. Since the advent of Chaos Theory, a new trick has been employed in areas as diverse as weather forecasting and banking: after the initial run, the computer repeats the procedure but with the original data slightly altered every time. If all the runs produce much the same result, we know the prediction is likely to be correct, but huge differences suggest that we may be in the presence of chaos.

In the early 1990s, I recall having a conversation with a journalist who has just returned from interviewing a pair of brilliant pure mathematicians whose expertise in Chaos Theory had just been rewarded with very well-paid jobs with an American bank. Impressed by their air of academic precision rather than the usual brashness of city high-fliers, the journalist asked whether their appointment signified that the era of the mathematical geek had finally arrived. 'We shall know that the era of the geek has arrived when Armani designs an anorak,' one of them replied.

If you type the word 'anorak' into the search box on www.armani.com, it comes up with 28 hits. So the age of the geek is definitely with us. During the opening of the new Tate Modern gallery in London in 2016, the fashion magazine *Vogue* reported: 'The outfit that stole the scene at last night's glamorous Tate Modern opening wasn't a sweeping silk slip dress, but Gucci's apple-green anorak. Love it or loathe it, high fashion's ambitions on geeky style are here to stay.'

And it may all be the result of Chaos Theory and that fluttering butterfly.

The development of Chaos Theory led to a widening of interest in unpredictable systems, leading in turn to the introduction of two further branches of mathematics which also, by pure coincidence, began with the letter C: Complexity Theory and Catastrophe Theory. Like *Chaos* and *The Butterfly Effect*, there was also a film released in 2011 with the title *Complexity*, and although I have found no reference to a film called 'Catastrophe', there were at least three TV series of that name in 2008, 2015 and 2017. None of those, however, bore more than a chance resemblance to the theories bearing the same names.

Neither Complexity Theory nor Catastrophe Theory has captured the public imagination to the extent that Chaos Theory has, but both are worth mentioning here, as they reflect different sides of our chaotic world. While Chaos Theory introduced the surprising concept of random behaviour emerging in an organized system, Complexity Theory is all about organization resulting from randomness, while Catastrophe Theory accounts for the sudden actions that may result from otherwise smooth changes.

Complexity Theory

With the ramifications of global trade, politics and the Internet, the world has become ever more connected and more complicated. Complexity Theory is an attempt to bundle together all such complications under one catch-all heading, but it is so complex that there is no general agreement on what it is. Complexity studies the order that may come from chaos.

Complexity Theory is, as befits its name, extremely complex. In fact, it is not so much a theory as an umbrella term for a set of ideas straddling the ways in which order and structure may emerge in apparently disorganized systems across various disciplines.

In biology, it may explain the graceful patterns of flocks of birds in flight resulting from unpredictable behaviour of individual birds. It has also been applied to the apparently coordinated behaviour of colonies of ants or bees, even though single insects are unpredictable. Similar organization has been shown to emerge among employees in large companies, though every employee may be doing his or her own thing.

In engineering, complexity may apply to the functioning of an entire assemblage such as an aeroplane made from millions of small components. In ecology, the functioning of an animal population or habitat of plants may be seen as a complex adaptation created by the behaviour of individual members. In economics, an entire financial system is similarly created by the decisions of individual investors.

The functioning of the human brain can also be seen as a system that emerges from the interactions of its individual cells, though we do not understand how, and a similar bemusement surrounds our attempts to predict the behaviour of socio-political systems worldwide.

In all such cases, we see global patterns on a grand scale arising from vast numbers of local interactions. When humans are involved, individual decisions may be affected by competition or cooperation, which in turn adapt, without any central control, according to their effect on the whole.

Complex Numbers

Perhaps confusingly, Complexity Theory has very little to do with complex numbers.

Complex numbers are a beautiful and surprisingly useful example of the ability of mathematicians to generalize their findings and their methods. We started with integers (whole numbers), with the concepts of zero and negative numbers added at different times in various cultures. Addition and multiplication led to division and fractional numbers, and multiplication led to the specific example of squaring a number, which proved very useful to Pythagoras when considering right-angled triangles, But squaring led to a difficult question about square roots:

We know that 2 and –2 are the square roots of 4, but what's the square root of –1?

Blinkered practical types may have said that negative numbers don't have square roots, but mathematicians did not want to exclude such equations as $x^2 = -1$ from consideration so they invented an 'imaginary' number, i, as the square root of minus 1. The term 'imaginary' was coined by René Descartes in the seventeenth century and was intended derisively, though such numbers were later found to be very useful.

Quite apart from giving us square roots of negative numbers, this imaginary i gave rise to complex numbers, which combine a 'real' part with an 'imaginary' part as $x + iy$. As if to prove their worth to Descartes, they provide a useful way of characterizing the points of a plane in Cartesian geometry. Descartes did not like the idea of complex numbers very much and even coined the term 'imaginary numbers' (by contrast with 'real numbers', which had no imaginary part) to express his disapproval, but he also saw the benefit of expressing the complex number $x + iy$ as the point with coordinates (x, y) in a plane.

Newton also tended to hold complex numbers in low regard but it was soon shown that any algebraic equation had solutions in complex numbers, even if it did not have an answer in real numbers. Although half-imaginary, complex numbers have applications in a wide range of mathematics and physics, including geometry, wave mechanics, fluid dynamics and quantum theory.

The pre-Newtonian scientific paradigm prescribed a predictable world. The post-Newtonian world of chaos and complexity was always in flux with no reassuring equilibrium points where everything settles down. It is too early to say whether Complexity Theory will itself settle down into a mathematical framework that is universally applicable to such complex systems and will at least let us define precisely what we mean by complexity.

Catastrophe Theory

Catastrophe Theory is a branch of mathematics first developed around 1970 to analyse sudden changes in dynamic systems such as avalanches, prison riots or changes in an animal's behaviour between flight and fight. While it provided a neat qualitative way of describing what was happening, limitations of the mathematics of Catastrophe Theory led many to doubt its applicability as a quantitative tool.

Catastrophe Theory, like Chaos Theory or Complexity Theory, covers a multitude of diverse topics, from the behaviour of dogs to the price of train tickets, but it is at least easier to understand and should not be confused with them. The word 'catastrophe', like 'chaos', comes to us from ancient Greek, but whereas chaos was the primeval state of disorganization before the world's creation, catastrophe was the sudden turn bringing about the end of a drama.

When the word began to be used in English in the late sixteenth century, it was, as the Greeks had used it, just as likely to be a happy ending as a disastrous one. Of the four catastrophes in the works of Shakespeare, one is a 'catastrophe of the old comedy' (in *King Lear*), another a catastrophe of a nuptial (*Love's Labour's Lost*), a third is unclear (in *All's Well That Ends Well*) and in *Henry IV Part 2*, it is used as a euphemism for buttocks in the phrase 'I'll tickle your catastrophe'.

As the centuries passed, catastrophes became increasingly identified with sudden disasters. When Catastrophe Theory was first used as the name of a branch of mathematics in 1971, however, it reverted to its original general meaning of an abrupt, discontinuous leap. Its basic idea is shown by a simple graph (see diagram on next page).

Let us suppose that the horizontal axis measures something within our control (such as proximity to a potentially vicious dog or the price of an article we produce) and the vertical axis is a measure of something we are trying to influence (the dog's aggression or our profits).

The portions of the curve from A to F and from C to D offer no problems: every value on the horizontal x-axis identifies just one point on the curve corresponding to a single value on the vertical y-axis. As the curve moves from F to C, however, every value on the x-axis corresponds to two values of y. Let's see what this means in practice as we gradually move along the curve starting at A.

As we proceed towards F, all is clear: the value on the

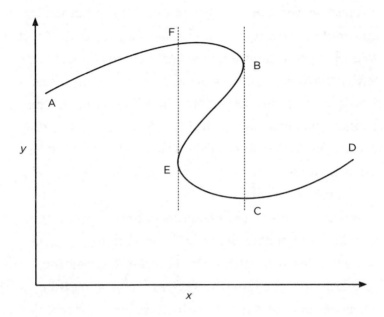

y-axis gradually increases, but after F it soon begins a slow decline. When it reaches B, the catastrophe happens: as x increases, it has nowhere to go and suddenly plummets to C. The mathematics gives rise to a double-valued function in the middle section with two y-values for every x-value, but, in practice, one of those values, when the curve doubles back on itself, is not realized. The result is a discontinuous curve and a catastrophic jump.

Some saw this as a possible economic model of the effects of continuous price rises. As prices slowly go up, they lead to higher profits but also affect sales. At a certain point after F on the graph, profits begin to decline, so prices are again raised, leading to a further decline in sales and a corresponding decline in profits. Finally, at point B, sales collapse as few can now afford the product and profits collapse to point C.

In the case of that worrying dog, the animal backs off as we get closer and closer, but there comes a point when it feels too threatened to continue retreating and suddenly attacks. This flight-or-fight response pattern had long been known, but catastrophe theory was the first to explain it mathematically.

The really imaginative part of Catastrophe Theory, however, lay in its representation of the origin of that winding, sometimes double-valued curve; to see the power of the theory, we need to introduce further dimensions. The next diagram is a simple three-dimensional example.

The z-axis measures what we are trying to control, but now this is influenced by both x and y, and these in turn may influence each other. So profit can depend on both price and sales, which of course affect each other. Or a dog's aggressive response may depend on both fear and anger. Now, instead of just a simple curve, we have an entire surface, shown by the broken lines. Again, we have the two-valued curve on the front face, but the back face may be a smooth single-valued curve. The entire surface is like a carpet with a fold in it, and the region of that fold gives rise to the characteristic possibility of catastrophe.

At its inception, Catastrophe Theory was seen as a possible mathematical solution to a wide variety of problems, but in the late 1970s doubts began to be raised about whether it truly modelled the real world, especially in economic fields.

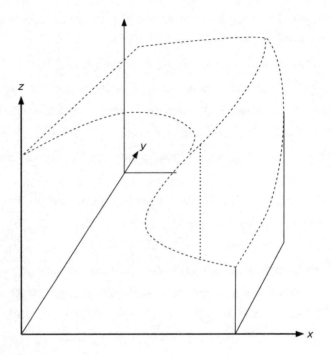

It has remained an elegantly descriptive mathematical theory, but the extent of its possible applications is unclear.

6	∞	9	3	8
2	7	=	1	%
4	¾	3	2	±
Σ	9	7	¼	3
¾	0	x	4	8

CHAPTER 11

Torpedoes, Toilets and True Love

A difficult problem with many applications

6	∞	9	3	8
2	7			%
4	¾			±
Σ	9			3
¾	0			8
3	√			+
6				9

Public Toilets should be the concern of every civilization because the cleanliness and standards of hygiene they do or do not set are truly a measure whereby the standards of a society can be gauged.

(Professor Wang Gung, at the opening of the 1995 International Symposium of Public Toilets in Hong Kong)

What, you may wonder, do toilets, torpedoes and true love have in common, other than the fact that they all begin with T? I could have included secretaries, spouses and supermarket queues for a different alliteration. What they all share is a mathematical theory that may apply to them all. Let's use the torpedo example to introduce it.

You are the captain of a wartime submarine and you are down to your last torpedo. The ships of the enemy navy are sailing past and you want to fire your torpedo at the biggest one. You can see them one by one as they pass and you can accurately judge their sizes, but you know nothing about the enemy flotilla other than the number of ships it has in total. Your only information about the sizes of the ships comes from seeing them as they sail past. For reasons of your own security, you cannot give away your position by moving, so if you decide not to fire at a particular ship as it passes, you cannot change your mind and go back to pursue it later. The question is: what strategy should you adopt to give the best chance of firing your torpedo at the biggest ship?

As a maths problem, this has been around since about 1950 but was brought to general attention ten years later when Martin Gardner wrote about it in his highly entertaining 'Mathematical Games' column in *Scientific American*. But his version was called 'The Secretary Problem' and had nothing to do with torpedoes. It was described as follows.

A manager has to appoint a secretary quickly and a certain number of applicants have applied for the job. He interviews them one by one, but when each interview is over, he cannot say 'we'll let you know' but must either make a job offer or send the candidate home, never to return. As with the submarine captain, the question is: how can he maximize his chance of picking the best candidate when all he knows is how many applicants there are?

Other versions of the problem include selecting the cleanest and most hygienic toilet at a music festival when you are at the front of a queue walking past them and examining them sequentially, or selecting a spouse (or finding your true love) from all the people you date. Again you cannot reject a toilet or potential spouse and expect it (or them) to be available later. Trying to pick the shortest checkout queue at a supermarket is another example, if you do not allow yourself to go back to a queue you have previously rejected.

In all cases, the best policy is to use the early candidates just as a source of information by which to assess the others. Once you have let the first few go past, you then set yourself the task of finding one that fits your criteria better than any you have previously seen.

To get back to the submarine captain, what he must do is let a few enemy ships sail past, then fire his torpedo at the first ship that is bigger than any of those previously seen. Before trying to work out the general optimal strategy, let's look at a specific case when the captain knows there are five ships in the enemy fleet. We'll call them 1, 2, 3, 4, 5 in order of size, but remember, they may sail past in any order.

Any of the five ships may sail past first and, whichever does so, there are four remaining ships that may be in second place. Whichever is second, there are three remaining that may be in third place, then two in fourth place, and finally the last ship. That gives a total of 120 ($5 \times 4 \times 3 \times 2 \times 1$) different possible orders of these ships, so let's try out a few strategies. To start, suppose we let just one ship go past and fire at anything bigger that follows. If the second biggest ship, which we have called 4, goes first, then we are on to a winner whatever happens: 5 is the only ship bigger, so that's the one we will fire at whatever order the remaining four ships come in. Having ship 4 in first position accounts for 24 ($4 \times 3 \times 2 \times 1$) of the 120 possible orders. We will also be happy if ship 5 is in second place behind 1, 2 or 3, which together cover another 18 possible orders (6 possible orders for each option). If ship 5 is in third place, however, we'll only get the desired result if the order of the first two ships is 2, 1 or 3, 2 or 3, 1 (we've already counted those beginning with 4) which together add another six to our total of good results, and finally, the only two other orders that work for our submarine captain are 3, 1, 2, 5, 4 and 3, 2, 1, 5, 4.

That gives a total of $24 + 18 + 6 + 2 = 50$ cases in which we succeed in hitting the largest ship. That, you will remember, was if we let only the first enemy ship sail past, but look at what happens if we let two go by instead. In that case, we'll miss our chance if 5 is in the first two, but the gains more than make up for this. In fact, we'll hit the biggest ship in 52 of the 120 possible cases, which is two better than before. However, if our submarine captain cautiously lets three ships pass before deploying his torpedo, he is successful in only 34 cases. So the optimal strategy here is to let two ships go past.

So let us see what we can work out in a general case of N ships of which we allow K to pass unhindered. The maths involved in working this out is far more difficult than anything we have previously seen in this book, but the following should give a good idea of how it is worked out.

There are two ways in which our strategy can be scuppered: either the biggest ship is included among the first K, or it is outside the first K but is preceded by another ship that is bigger than anything in the first K. (For example, the order 3, 1, 2, 4, 5, for $K = 1$, 2 or 3. In those cases, we will choose 4 before we get to 5.) So let's try to calculate our chances of success.

Since the ships are assumed to pass in random order, the biggest ship is equally likely to be in any one of the N places, so the chance of its being in any particular place is $1/N$. That's the easy part; now comes something that took me ages to understand.

As we said, if the biggest ship is in any of the first K places, we will sadly let it go past, but what happens if it is in position $K + i$, for any value of i greater than zero up to $N - K$ (when $K + i = N$ we arrive at the final ship)? The first such case is $i = 1$, which is easy: if the biggest ship is in place $K + 1$, we see it immediately after ignoring the first K ships. As it is the biggest ship, it is bigger than any of the first K, so we fire our torpedo. Higher values of i are a bit trickier. In such cases, we only fire at the biggest ship if we have not fired the torpedo at any of the ships in places between $K + 1$ and $K + i - 1$ inclusive. So let us turn our attention to the first $K + i - 1$ ships. If the biggest of those ships is in the first K, we are fine. We shall continue to ignore the ships that pass until we get to $K + i$, but if it is among the ships between $K + 1$ and $K + i - 1$, then we will fire our torpedo at it and never reach our biggest ship at $K + i$.

Now just as the chance of the biggest ship overall being in any specific position is $1/N$, the biggest ship among the first

$K + i - 1$ ships has an equal chance of being in any of those $K + i - 1$ positions, which is equal to $1/(K + i - 1)$. The chance of its being in the first K places (which is what we want) is therefore $K/(K + i - 1)$, while the chance we will have wasted our torpedo before we get to the largest ship at $K + i$ is equal to $(i - 1)/(K + i - 1)$.

If we now multiply $K/(K + i - 1)$ by $1/N$, we get the chance of the biggest ship being in place $K + i$, which is $1/N$, and our firing at it, which is $K/(K + i - 1)$. Finally, we must add together those answers for every value of i from 1 to $N - K$.

Since every term of that sum contains the factors $1/N$ and K, we are left with a fairly simple equation for the probability P of achieving our goal:

$$P = \frac{K}{N}\left(\frac{1}{K} + \frac{1}{K+1} + \frac{1}{K+2} + ... + \frac{1}{N-1}\right)$$

In the simple example given above with $N = 5$ and $K = 2$, we get

$$P = \frac{2}{5}\left(\frac{1}{2} + \frac{1}{3} + \frac{1}{4}\right) = \frac{2}{5}\left(\frac{13}{12}\right) = \frac{26}{60}$$

which is equal to the figure of $52/120$ that we obtained by counting.

The real question, however, is to determine the value of K which maximizes P for any given N. In other words, how many ships, or supermarket queues, or toilets should we let pass before we start committing ourselves to picking one. The answer to that needs some rather complex (but elegant) maths involving calculus, so you have my full permission to skip it, though the conclusion is well worth taking note of and even provides some reassuring evidence that people are more mathematical than they may appear.

So let's get back to that formula:

$$P = \frac{K}{N}\left(\frac{1}{K} + \frac{1}{K+1} + \frac{1}{K+2} + \ldots + \frac{1}{N-1}\right)$$

It turns out (I will skip the proof as it is too technical for this book) that the part inside the brackets is known as an approximation for the value of $\log(N/K)$ – and here (please don't be put off) we are talking natural logs to base e rather than the sort of logs you may have seen in log tables at school, which are to base 10 – and the approximation becomes better and better as we increase N while keeping the ratio K/N the same. So, applying the approximation to the formula, we now have

$$P = \frac{K}{N}\log\left(\frac{N}{K}\right)$$

or, to tidy it up still further,

$$P = x\log\left(\frac{1}{x}\right), \text{ where } x = \frac{K}{N}$$

e and Transcendence

Of all the numbers occurring naturally in the mathematical jungle, two stand out above all others. Everyone knows about the ratio of the circumference of a circle to its diameter, which we know as π (pronounced pi and named after the Greek letter that starts the word *perimetros*, meaning perimeter). But the other highly significant number, while no less important, is far less well known.

This number, which we call e, crops up all over the place in calculus and in wave theory, including light, sound and quantum mechanics, but most of all in probability theory and the calculation of compound interest. But there is no example of e in use that is as simple as the ratio of a circle's circumference to its diameter, so e plays a backstage role as pi's shy cousin.

Perhaps the simplest example of e is the way it made its first appearance when it was discovered by the great Swiss mathematician Leonhard Euler around 1748. The problem he was working on at the time concerned compound interest. Suppose you invest £1 at 100% interest a year. Then you will receive £1 interest at the end of the year, giving a total of £2, but how does this compare with getting 50% interest every six months, or 25% interest every three months, each time adding the interest to our capital and reinvesting the total? What if the interest is paid daily, or every hour, or every minute?

In each case, if the interest is paid in n equal intervals at a percentage rate of $\left(100+\frac{100}{n}\right)$ each time and the interest then reinvested, our total at the end of the year works out at $\left(1+\frac{1}{n}\right)^n$.

For $n = 2$ (which is six-monthly payment), that gives a total of £2.25 at the end of the year, which is equivalent to 125% interest on our capital. For $n = 4$, after a year our £1 has grown to £$(1.25)^4$, which is £2.44 (to the nearest penny).

The more often interest is paid, the more the number grows, but Euler made the remarkable discovery that, as the value of n increases, $\left(1+\frac{1}{n}\right)^n$ does not grow arbitrarily large but approaches a limit. For $n = 100$, its value is 2.705, but by the time we get to $n = 1,000$ the value of $\left(1+\frac{1}{n}\right)^n$ has only grown to 2.717 and it has only just crept up to 2.718 for $n = 10,000$ or even $n = 100,000$. The value of this limit is what Euler gave the name e.

On the subject of e and π, we have mentioned before that numbers come in several varieties, each a little weirder than the one that came before. First there are the integers: 1, 2, 3, 4, ... and the negative integers, −1, −2, −3, ... Zero is also usually considered an integer, though some think its nothingness does not deserve such an accolade.

Then come the rational numbers, or fractions as we sometimes call them, such as ⅔, ²²⁄₇, ¹⁄₁₀,₀₀₀, which are expressed as a ratio between two integers.

Next, we have irrational numbers, which cannot be expressed as fractions and may be *algebraic* or *transcendental*.

Algebraic numbers are numbers such as √2 which cannot be expressed as ratios between two integers. Any number that can satisfy an algebraic equation, such as $x^2 = 2$ or $x^5 = 7$ or $x^5 − x = 1$ is an algebraic number. And this brings us to the subject of irrationality and transcendence.

The phrase 'irrational number' sounds like an oxymoron. After all, what is more rational than numbers? The irrationality of such numbers, however, lies not in illogicality but in the fact that they cannot be expressed as a ratio between two other numbers.

The Pythagorean philosophers, around 500 BC, believed that rational numbers were the key to everything and were really upset when the philosopher Hippasus discovered that the square root of 2 was not rational. According to legend, he was murdered to keep that fact a secret. But the proof that $\sqrt{2}$ is not rational is fairly simple.

Suppose $\sqrt{2}$ can be expressed as a rational number which we shall call a/b (expressed in its lowest terms, so a and b are integers with no common factor). Then $\left(\frac{a}{b}\right)^2 = 2$, which means that $a^2 = 2b^2$. So a^2 is divisible by 2, which means that a is an even number (as the square of any odd number is another odd number).

So we can say $a = 2c$ where c is another integer, which means that $a^2 = 4c^2$.

But $a^2 = 2b^2$ so $2b^2 = 4c^2$, which means that $b^2 = 2c^2$, so b must also be divisible by 2, just as we proved earlier that a was. But that contradicts our earlier statement that a and b had no common factor, so a/b cannot be the square root of 2. QED.

> Most exotic of all are the transcendental numbers, which are numbers such as e and π that cannot be expressed as solutions to an algebraic equation. However, it is very difficult to prove that any particular number is transcendental. In fact it was not until 1873 that the French mathematician Charles Hermite proved that e was transcendental, and the transcendence of π was first proved by the German mathematician Ferdinand von Lindemann in 1882.

When we started all this, our object was to find the value of K for a given N that optimized our chances, in other words gave the highest possible value to P. Now at last, with a little knowledge of calculus, we can do precisely that, and the answer that emerges is that, for any N, we should choose the value for K that is closest to making K/N equal to $1/e$.

Yes, it's that same elusive e again, and its value is approximately 2.718. So that's the answer: we should reject the first $1/2.718$ (which is 37%) of the ships, supermarket queues, potential spouses, toilets or secretaries, then pick the first one we see that is better than any of those we rejected. And it turns out that if we use this strategy, our chance of picking the best one of all is also $1/e$. If you cannot recall the value of e, remember to let just over a third of the candidates at the start go past and you will not go far wrong.

For many of us who do not go to music festivals or command submarines, perhaps the most interesting application of this algorithm concerns the matter of spouse selection, and that is where we see what good mathematicians we all are.

Algorithm

An algorithm is essentially nothing more than a step-by-step set of instructions to be followed in order to reach a desired objective or get a desired result. Any cooking recipe is an algorithm; any decent flat-pack assembly instructions are an algorithm; any computer program is an algorithm.

As computer pioneer Alan Turing himself pointed out, however, computers are restricted by having to use algorithms. There are several aspects of human thought, such as creativity and consciousness, that do not seem to be the result of algorithmic processes in the brain. Producing computer programs to simulate these is one of the greatest challenges for artificial intelligence.

The word 'algorithm', incidentally, came from the Latinization of the name of the eighth-century Persian mathematician Muhammad ibn Musa al-Khwarizmi, who was a member of the House of Wisdom in Baghdad.

According to the above analysis, we should reject just over a third of all possible spouses, then choose the first one we see who is better than any seen during that sampling phase. There is a problem here because we do not know how many potential spouses we will ever meet, so we cannot really judge how many would constitute a third, but we could instead make the calculation based on time. Most people actively look for a spouse between the ages of 19 and 40, which gives a period of 21 years to find the best possible candidate. Mathematically, the best strategy would therefore be to spend the first $^{21}/_{2.718}$ years assessing potential candidates, then pick the first who is better than any of those seen in the sampling period. That sampling period works out at about 7.8 years, so according to this analysis, the best chance of finding the perfect spouse, assuming you start dating at the age of 19, would be to wait until you are 26.8 before even thinking about making a decision.

A recent survey in the UK reported that the average married couple spent 4.9 years dating, so if your search does not become serious until you are 26.8, you will need 4.9 years after that, which takes us up to the age of 31.7. Amazingly, the average age for women in the UK to marry (for the first time) is 31.5, while for men it is 33.4. That either means that women come closer to employing the perfect strategy or that men take almost two years longer after their sampling period before meeting Mrs Right, whom they then marry after 4.9 years of dating. Both male and female figures are close enough to the mathematical prediction to suggest that, on average, people are following the best strategy.

Interestingly, the mathematics tells us that the best possible strategy only gives the best possible result 37% of the time, which suggests that 63% of marriages are sub-optimal, yet according to the UK Office for National Statistics, only 42% of marriages end in divorce. Of course sub-optimality has never been formally accepted as grounds for divorce.

6	∞	9	3	8
2	7	=	1	%
4	¾	3	2	±
Σ	9	7	¼	3
¾	0	x	4	8

CHAPTER 12

Formula Milking

Unbelievable formulae in newspapers

Some formulas are too complex and I don't want anything to do with them.

(Bob Dylan, interview, 2009)

Newspapers love formulae. They look erudite; they are eye-catching on the page; and most readers do not understand them so are suitably impressed. They are also often meaningless, or incorrect, or are misrepresented to the point of becoming gibberish.

Entering the phrase 'formula for the perfect' into the search bar for Google news pages results in over 26,000 hits. There is a formula for the perfect cup of tea or coffee, the perfect sports bra, the perfect mince pie, the perfect Christmas song, the perfect cheese sandwich, the perfect holiday, the perfect pancake toss and the perfect biscuit dunking. Despite the claim to perfection, they all leave a great deal to be desired. Yet still they appear, mainly because

PR agencies know that such things are eagerly lapped up by journalists and newspaper editors.

Equations and Formulae

In the closing days of 2019, I read one news report asking whether an 'Eastern Mediterranean equation was possible without Turkey'. Meanwhile, an American football report told us that in Nebraska 'tight ends need to be part of the equation' while back in New York, another newspaper gave its opinion on 'How a strong economy affects the impeachment equation'.

'Equation' has become a word beloved of journalists in order to add spurious mathematical validity, yet strictly speaking none of the above is an equation. An equation is a mathematical statement consisting of two expressions with an equals sign between them. For example,

$$(x + y)^2 = x^2 + 2xy + y^2$$

is an equation. So is Einstein's energy equation:

$$E = mc^2$$

But the Eastern Mediterranean, football tactics and impeachment are not equations. At a stretch, these could perhaps be described as formulae, as this term

---

conveys a different meaning and purpose.

A formula is a mathematical rule, usually expressed in symbols. It is a way of calculating something you don't know from values you can measure.

Einstein's equation $E = mc^2$ may also be seen as a formula for calculating energy, E, from mass m and the speed of light c.

$F = \frac{9C}{5} + 32$ is a formula for calculating the Fahrenheit equivalent of a Celsius temperature.

A formula is a mathematical algorithm. It is also a specific type of equation but one in which one side of the equals sign is a single symbol signifying what we are trying to work out.

Here are a few more examples of formulae:

$V = \frac{1}{3}\pi r^2 h$ gives the volume V of a cone in terms of its height h and radius of its base r.

$A = 4\pi r^2$ gives the area A of a sphere in terms of its radius r.

$x = \dfrac{-b \pm \sqrt{(b^2 - 4ac)}}{2a}$ is the formula to calculate the solutions to the quadratic equation $ax^2 + bx + c = 0$.

I only give these definitions and examples because of the number of times I have seen a formula described as an 'equation'. Technically it isn't wrong, but it gives an incorrect impression of its purpose. An equation asserts that the two expressions on either side of the equals sign are equal and that they carry equal weight in our minds. A formula is specifically a means for calculating one thing from others. Formulae can be described as equations, but not all equations are formulae.

I must confess my guilt in producing some of these formulae myself, but in my defence I use reputable statistical techniques rather than just conjuring up some spurious mathematical symbols that will impress PR people.

I have already mentioned (page 42) my efforts in producing a formula for female beauty, but another favourite personal formulaic endeavour arose when a betting firm asked me if I could produce a mathematical formula to help people pick a winner in their office Grand National lottery. I pointed out that randomness is an essential feature of lotteries, so mathematics was not going to help, but I offered to see what mathematics could do to help predict which horse would win.

Unhampered by any knowledge of horseracing, I proceeded to analyse the names of winning horses over the entire long history of the Grand National race and noted down how many words were in the name, how many letters, and what letter of the alphabet the name began with. Remarkably, I found that more winners' names began with the letter R than any other letter, with the next favourite first letters being A, S, M, and horses with one word in their name tended to do better than those with two, three or more names. Equally remarkably, the winners frequently had eight or ten letters in their name, but only rarely had nine.

The only other quantifiable information I had on the horses was their age in years, so I devised a scoring system based on the first letter, number of words and letters in the name and the horse's age in years, allocating up to four points for each criterion. This produced a score of up to 16 for every horse, thus enabling me to pick the horses which, according to my system, were most likely to win.

Almost every British national newspaper carried stories about my analysis, and a friend was so impressed that he put small bets on each of my top three tips. I was delighted to hear that he won a total of nine pence.

It is, of course, very unlikely that a horse's name will play a great part, or indeed any part at all, in its success but my system did at least have some authenticity. The entire science of psychometrics is based on the idea of devising tests that measure a variety of aspects of intelligence or personality, then seeing whether the tests can discriminate between groups consisting of those who are good and bad at whatever one is trying to select for.

When a test does show such ability, a proper regression analysis will even produce a formula that picks the best people in the most effective manner. I had done much the same thing with the horses' names. To do it properly, I should really have used only half my sample of winning horses to derive the scoring system, then tried to validate it by seeing if it applied to the other half too, but that, I fear, would have spoiled the fun.

While my formula might not have been very successful at picking winning horses, it did at least provide clear rules to follow in working out scores. When a newspaper publishes a 'formula', it often does not even do that, but it always credits the gibberish it publishes to a 'professor of pure mathematics' or a 'top psychologist' and usually includes the words 'secret' and 'revealed'.

The 'formula' may be nothing more than a list of recommended ingredients, as in a 'formula for the perfect shopping trip' published in 2016 which was really nothing more than a list of 20 desirable features such as 'grabbing a bargain', 'being able to find my way around easily' and 'loving the first thing I try on'.

Or it may be a ridiculous simplification of an intractable question, such as the 'formula for the perfect joke' revealed in 2008:

$$x = \frac{(fl + no)}{p}$$

which was alleged to measure the excellence x of a joke as the funniness (f) of the punchline multiplied by the length of build-up (*l*), to which is added the amount someone falls down (*n*) multiplied by the physical pain or social embarrassment the joke causes (*o* for 'ouch') and the total divided by the punning factor (*p*), which supposedly reduces laughter and produces more of a groan. No explanation was offered about how to assign numbers to any of those elements or what score one should be aiming for.

Another over-simplified formula that grabbed media attention was the 'formula for a perfect flight', which was widely published in 2015, offering the simple calculation:

$$F = (T + L - 30)\frac{P}{100}$$

giving *F*, the excellence of the flight, in terms of *T* (time of day), *L* (legroom) and *P* (punctuality). At least this article explained clearly how to calculate the various elements: for the time of day, score 10 for morning flights, 5 for night flights and 3 for afternoon flights; *L* is the legroom in inches (ranging from 28 to 40) and *P* (for punctuality) is the percentage of that airline's flights that arrive on time. The maximum value $T + L - 30$ is therefore 20, and $P/100$ has a maximum of 1, so the value of *F* will be between 0 and 20. A good flight, we were told, will score 15 or more, while a perfect flight scores 20. Sadly, the formula does not take account of the qualities of the person you are seated next to, or the in-flight food, or the helpfulness of the crew, or the rowdiness of the other passengers or countless other things that may contribute to our assessment of the excellence of the flight.

Sometimes, the formula will go to the other extreme, looking so complicated that it becomes unintelligible. In 2012, more than one newspaper in the UK published a 'formula for the perfect holiday', which looked like this:

$$((N(d)\mu(d)-40)(r))/(\sigma(b)((C(d)-\mu(d))N(d)-41/40c))(41c.a)/[40]^2$$

That formula may have made sense when it left the mathematician who devised it for the hotel company that commissioned him, but it had clearly turned to nonsense by the time it appeared in the newspaper. Even telling us that $N(d)$ was the number of possible holidays of length d that can be taken in a year and $C(d)$ is the cost of that holiday dependent on the number of days taken and $\mu(d)$ refers to the holidaymaker's anxiety levels does not help much, and all the other variables are similarly unhelpfully explained.

In 2004, Giles Wilson wrote a fine piece about such formulae in the BBC News Online Magazine. After giving some examples of the glut of dodgy mathematics in newspapers, he suggested the following:

$$H = O\,(f+\mu) + S$$

as the 'formula for the perfect formula', from which one may calculate H, the number and prominence of the headlines any formula generates, in terms of O (the ordinariness of human behaviour you're explaining), f (an unexplained factor for having a formula worked out), μ (the presence of a suitably scientific-looking symbol) and S (having a sponsor with an enterprising public relations office).

While agreeing with this in principle, I would suggest that the $(f + \mu)$ factor could be elaborated with f dependent on the length of the formula and number of variables and μ explained at greater length, with integral signs, square roots, Greek letters and power exponents all gaining extra points.

Apart from that formula for the perfect formula, I think perhaps the formula offering greatest promise was one produced by Manchester Metropolitan University in 2006 supposedly giving a way to calculate the perfect bum:

$$\frac{(S+C)(B+F)}{(T-V)}$$

where S is shape, C is circularity, B is bounciness, F is firmness, T is skin texture and V is the ratio of hips to waist. The first five of these are assessed on a scale of 0 to 20, and the nearer the formula gives to a value of 80, the more perfect the bum is supposed to be.

More than a decade after this appeared, however, a paper in the *Journal of Plastic and Reconstructive Surgery* came up with a much simpler answer. The ratio of waist size to hips, it said, should be 0.7 – and that, I feel, should be accepted as the bottom line.

6	∞	9	3	8
2	7	=	1	%
4	¾	3	2	±
Σ	9	7	¼	3
¾	0	x	4	8

CHAPTER 13

Monkey Maths

An evolutionary perspective on numeracy

Σ	9	7	¼	3
¾	0			8
3	√			+
6	4			9
¼	0			7
5	%			3
0				½

Some Americans I have spoken with, (who were otherwise of quick and rational parts enough) could not, as we do, by any means count to one thousand; nor had any distinct idea of that number, though they could reckon very well to twenty.

(John Locke, *An Essay Concerning Human Understanding Part 1*, 1689)

The earlier chapters in this book were all about various aspects of human innumeracy or irrationality. Whether we are failing to grasp the complexities of Chaos Theory or making mistakes involving simple percentages, the majority of us are constantly tripping up over numbers. As several anthropologists have found, however, some societies do not suffer from such problems as they scarcely have any numbers at all.

As John Locke pointed out, writing about a small tribe in Brazil: 'The Tououpinambos had no names for numbers above five; any number beyond that they made out by showing their fingers, and the fingers of others who were present' and such numerical 'inadequacy' was by no means unique among peoples that we may once have considered primitive.

Daniel Everett, writing in 1986 about the Pirahã people of Brazil, said that they had words for one and two, but no higher numbers. Later he revised that view, having decided that their language had no number words at all and the words he had referred to actually meant 'small quantity' and 'larger quantity'. Attempts to teach them to count have not met with great success, largely because the concept of numbers does not interest them in the slightest. They are not so much innumerate as anumerate, having no use for precise counting at all. In fact, they have a word for all languages other than their own, which is a term meaning 'crooked head', indicating the laughable inferiority in which they hold other cultures that waste their time counting things.

Until recently, the abstraction of counting was seen as a mark of human intelligence and perhaps even an important aspect of our evolutionary development. Numbers, after all, give us the ability to make precise calculations, which must make us fitter for survival. Recently, however, several studies have identified counting ability in a wide variety of animals.

Salamanders and frogs, for example, have shown the ability to distinguish between one and two objects, or between two and three objects, though they have problems telling the difference between three and four. An experiment with bees, on the other hand, in which they had to choose between tunnels marked with different numbers of dots, showed that they can count up to five, while honey bees in particular have been shown to understand the concept of zero. Interestingly, bees have been seen to be faster learners if they are punished for getting things wrong.

Horses, spiders, bears, lions, crows and even three-day-old chicks have all shown some counting ability, as have our closer relatives, macaques and chimpanzees. It was experiments on capuchin monkeys conducted around 2005, however, that provided the most startling results.

Keith Chen, an economist from Yale University, and Laurie Santos, a psychology professor at the same university, knew that their research animals were clever but wondered if they were bright enough to understand money. So they created a supply of silver tokens with holes in the middle and taught the monkeys that their tokens could be exchanged for pieces of fruit. This took some time as the concept was strange to the capuchins, but once they had got the hang of it, their economic skills could be tested.

For the first experiments, the monkeys were each given a handful of coins before they were set loose in a 'market' where various researchers, wearing clothing that made them easy to distinguish from one another, offered different plates of food which could be bought for the tokens. Very quickly, the monkeys got the hang of money, allowing more detailed experiments to begin and proper observations to be made. One of the first of those observations was that the monkeys spent all the money they were given, and they even stole tokens that they found lying around, but it was the details of how they spent the money that was most interesting.

Early experiments established the value the monkeys assigned to portions of grapes, apples and jelly, specifically seeing how much of each item any particular monkey would buy for one money disc. Once the value of each item had been established and successfully communicated to the capuchins, the monkeys were given 12 discs and allowed into the market to buy whichever plates of foodstuff they preferred.

Then the experimenters, as any market traders are liable to do, introduced a special offer, effectively halving the prices of apples by doubling the size of apple portions. At the same time, the number of discs given to each monkey was dropped from 12 to 9. The amazing result was that apple consumption rose in exactly the way that economists' price theory (as applied to humans) would predict. Indeed, measured over the course of ten sessions, monkey consumption was within 1% of the theory's prediction.

Three cheers for the capuchins, but even more surprising results were to follow when Chen and Santos set the monkeys a pair of problems devised by Amos Tversky and Daniel Kahneman which had demonstrated irrationality in human decision-making. Here are the human versions, which are similar to some of the problems given in Chapter 3:

> **Problem 1. You are given $1,000 and offered the option of tossing a coin. If it comes down heads, you will be given another $1,000; if it comes down tails, you get nothing extra. Or you can reject the coin-tossing option and just get an extra $500.**

In the first case, you go away with either $2,000 or $1,000; in the second case, you will simply get $1,500, so on average they are both worth $1,500.

When the problem is posed this way, a clear majority of people choose to gamble on the coin toss.

> **Problem 2. You are given $2,000 and again offered the option of tossing a coin, but this time if it comes down heads, you lose $1,000 and if it is tails, you keep the $2,000.**
>
> **Or you can just pay $500 not to take the risk of tossing the coin and losing twice that amount.**

When presented with the choice that way, the majority of people will hand over the $500 to avoid the risk.

The two problems, of course, offer exactly the same choices: in the first case, you get either $2,000 or $1,000 and, in the second case, you secure a safe $1,500 without gambling, but people's choices vary according to the way the potential rewards are presented to them.

Tversky and Kahneman coined the term 'loss aversion' to explain the apparently illogical difference in results. In Problem 1, people saw the coin toss as an option to gain a bonus; in Problem 2, they saw it as risking a loss.

So what would the capuchin monkeys make of such a situation? They had already shown that they were good economists but would they be good enough to avoid displaying human irrationality? Experiments to find the answer to that question were designed, giving a monkey the choice between two salesmen.

First, the monkey had to choose between a salesman who sold one piece of apple for a disc, and a second salesman who offered two pieces of apple for a disc, but in half the transactions handed over only one piece. Despite this blatant fraud, the monkeys quickly saw that the second salesman offered a better overall deal, so came to prefer him.

Then the salesmen changed their behaviour. The second salesman would still offer two pieces and sometimes hand over only one, but now the first salesman added a bonus piece in half of the transactions. The average outcome was thus one and a half pieces of apple with either salesman, but the monkeys quickly changed their preference, preferring the first salesman, who sometimes added to the portion, to the second salesman, who sometimes took something away.

In the next version, the first salesman, as previously, displayed one piece of apple but now never gave a bonus, while the second salesman displayed two pieces but always took one away before handing them over. Again, the outcomes were identical, but this time the monkeys preferred the first salesman even more strongly, even reacting aggressively to the second salesman's behaviour. As one experimenter commented, the monkeys behaved in exactly the same manner as humans, except that human subjects did not generally throw their poop at the experimenters.

In recent decades, both philosophers and psychologists have come to favour the view that human decision-making is guided by two separate processes, one rational and one emotional. Our instant reaction is guided primarily by emotions before our analytic side thinks things over and comes up with a reasoned assessment. When these two processes reach different conclusions, it's the emotional one that wins more often than not, partly because confirmation bias starts acting as soon as the emotional reaction sets in.

Early in primate evolution, fast and powerful emotional reactions allied to risk aversion may have had strong survival value, enabling our distant ancestors to perceive threats and react quickly to them. Present-day economic risk aversion in humans and similar risk aversion in monkeys (accompanied by faeces throwing) may both have their origins in our shared evolution.

CHAPTER 14

Pandemic Pandemonium

The world's reaction to coronavirus

People are not so interested in the numbers themselves – they want to say why they are so high, and ascribe blame.

(David Spiegelhalter, 'Coronavirus Deaths:
How Does Britain Compare with Other Countries?',
Guardian, 30 April 2020)

In 2020, a deluge of numbers rained down on the world to an extent never seen before, describing the effects of a new strain of coronavirus and the associated disease Covid-19, which was causing turmoil in every country on Earth. The numbers related to the daily count of those infected by the disease and those who had died from the disease.

While people tried to make sense of the numbers, politicians faced the task of having to take decisions. The figures showed that the disease was spreading at an alarming rate and was causing increasing fatalities, but estimating those rates was not easy. The development of any epidemic will in general follow a precise statistical path, but, as we shall see, the mathematics that describes that path can only be worked out in hindsight, and hindsight comes too late to make the right decisions.

Making difficult decisions was not the only problem for the politicians: they also had to join forces with journalists to explain to the public what was happening. This resulted in a crash course for everyone in the statistics of epidemiology, and one of the first casualties in that crash was a very important word: 'exponential'.

Growth Rates

There are essentially two distinct ways in which something can grow: 'linear' and 'exponential'.

When something increases by the same amount in equal time periods, this is called linear growth. Our hair grows at a rate of about 1.25 cm (half an inch) a month; our fingernails grow at a rate of about 3.5 mm (0.14 inches) a month; our toenails can only manage about half that rate. Overall length increases at a constant rate until we cut our hair or nails, or they fall out or break. If we plot their values on a graph, it results in a straight line, which is why it is called 'linear growth'.

When something increases not by the same amount in each given time period but by the same proportion, the growth is called exponential, and it is much more interesting. If you draw a graph of exponential growth, it gives a curve that becomes steeper and steeper. However slow an exponential growth rate may be, it will always overtake any linear rate if it is maintained for long enough.

There is another type of growth called 'geometric', which is only slightly different from exponential. The technical difference is that 'geometric growth' is used for an increase that is measured in discrete jumps while 'exponential growth' is continuous. For all practical purposes, however, the two terms are almost interchangeable. We hear far more about exponential growth, because 'exponential' sounds a snazzier word.

Even before Covid came along, 'exponential' had become a buzzword when making predictions of huge growth rates. As 2019 drew to a close, there were optimistic forecasts of exponential growth in sales of video surveillance equipment and the adult trampoline market, among others, and alarmist reports of exponential world population growth and US immigration figures, but they all seemed to be based on a misunderstanding of what exponential growth really means. In particular, several reports included a phrase predicting 'exponential growth by 2025' or some similar date. As we shall see, that's always a warning sign that the writer may not understand the term 'exponential'. It is not just a synonym for 'huge', or 'doubling in a short time', though one would hardly guess it from the frequent misuses, particularly when referring to the Covid epidemic.

Exponential growth may start very small but its growth by a constant fraction of the total over each time period ensures that whatever we are measuring grows very large if you wait long enough. That's why 'exponential growth by 2025' is meaningless. Of its very nature, exponential growth is shown over a period of time, not at a certain moment. While linear growth describes growth by the same amount each time, such as 1, 2, 3, 4, 5, exponential growth may be 1, 2, 4, 8, 16, where each successive term is twice the previous one, or it may be 1, 1.1, 1.21, 1.331, 1.4641, etc., where each term is 1.1 times the previous one. This, incidentally, takes 49 terms before it overtakes our linear 1, 2, 3, 4, ... series.

Any increasing growth, however, is liable to be limited by finite resources or available market, particularly exponential growth with a high factor. Take the growth in numbers of users of the Internet worldwide in the 1990s, for example. Table 1 gives the numbers in millions at the end of each year and the growth factor (GF) compared with the previous year.

Table 1

Year	Internet users	GF
1995	16	---
1996	36	2.25
1997	70	1.94
1998	147	2.10
1999	248	1.69

The figures display something close to exponential doubling every year for the first few years, but then the growth rate drops to 1.69, and the following year it was down to 1.46. Two years later, the limiting effect of market saturation had reduced the growth rate to 1.13, and it stayed between 1.09 and 1.25 before hitting an all-time low of 1.04 between 2017 and 2018, by which time almost 60% of the world population had Internet access.

World population itself also showed exponential growth for some time and then fell off. Table 2 gives the years in which world population passed certain milestones.

Table 2

World population	Year
1 billion	1800
2 billion	1927
3 billion	1960
4 billion	1974
5 billion	1987
6 billion	1999

7 billion	2012
8 billion	2023 (estimated)

As the table shows, it took 127 years for the population to double from 1 billion to 2 billion, but the next doubling, to 4 billion, took only another 47 years, and the population is expected to reach 8 billion 49 years after that. These last two figures are very similar, suggesting an exponential growth rate, but compared with the earlier slow doubling, the rate has clearly slowed down. In fact, if we look at the annual increase in recent years, expressed as a percentage of the total population, we find that the rate has declined from around 2.2% in the 1960s to only 1.1% now and it is still falling.

Looking at the last six figures in Table 2, we can quickly calculate the number of years the world population took to reach each milestone from 3 billion to 8 billion. These are 14, 13, 12, 13 and 11 years, which strongly suggests an almost linear rate of increase.

The main reasons for the rapid increase in the 1950s and 1960s were improved agricultural productivity and medical advances, particularly in their success in greatly reducing child mortality. This last factor, together with the introduction of contraception, led to a decrease in human fertility rates from an average of 5 children per woman in 1950 to less than 2.5 today.

These examples should remind us, as we struggle to interpret the coronavirus figures, that the numbers alone are rarely enough to draw a proper picture of what is happening: we also need to examine the reasons behind those figures.

Table 3 shows the weekly death rates in UK hospitals of patients who had been diagnosed with coronavirus in the early weeks of the pandemic and gives a good example of something that isn't quite exponential growth.

Table 3

Dates	Deaths	% increase
5–11 March	7	----
12–18 March	115	1,543%
19–25 March	694	503%
26 Mar–1 Apr	3,095	346%
2–8 April	8,505	175%
9–15 April	14,915	75%
16–22 April	21,060	41%
23–29 April	26,097	24%

The early daily figures, however, carried a clear threat of exponential growth. Until 13 March, there were only one or two reported deaths on any particular day, but this shot up to 10 on 14 March, 20 on 16 March, 40 on 19 March, 87 on 24 March, 181 on 27 March, 374 on 30 March and 670 on 1 April. These figures show the number of days the figures took to double were 2, 3, 5, 3, 3, 2 which was close to an exponential rate of doubling every three days.

The next doubling, however, never happened. On 23 March, the UK government imposed a lockdown on the country, banning all large gatherings, closing all restaurants and places of entertainment, asking people to stay in their own homes as far as possible and at all times to abide by socially isolating measures designed to restrict drastically chances of transmission of the disease.

Similar measures were adopted by most other countries affected by the coronavirus, though the severity with which they were enforced varied considerably. While the British police approached groups of people with a gentle word to suggest that, for their own good, they should go home, South Africa employed 70,000 extra troops, ready to impose heavy fines or prison sentences on anyone breaking lockdown restrictions.

At the peak of the pandemic, over half the world's population were in lockdown, and it seemed to work. One of the few things known about the disease in its early months was that, for those most severely affected, the period from infection to death was between four and five weeks so the effectiveness of the lockdown could be assessed only after that period had elapsed. Sure enough, after that time, figures for both confirmed infections and deaths had begun to fall, but many questions remained unanswered.

Was the economic damage caused by the lockdown even worse than the suffering it was designed to prevent, in either the short term or the long term? What would have happened if considerably less stringent restrictions had been adopted, perhaps affecting only known carriers or sectors of society known to be most vulnerable to the disease?

The trouble was that scarcely anything like this had happened before. Even the pandemic of Spanish influenza in 1918–19 had not caused such disruption, despite affecting a third of the world's population – around 500 million people – and causing 50 million deaths. Sporadic attempts at quarantine or isolation were made, but the pandemic began during the First World War, and the war itself demanded far greater attention.

The British, however, had suffered one remarkably similar incident more than three and a half centuries earlier as the graph shows.

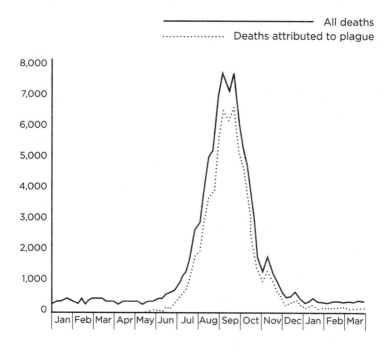

Mortality in the Great Plague, London, 1665–6

The pattern it shows is very similar to those of the coronavirus epidemic: a slow start for an extended period, then a rapid rise to a peak, followed by a decline almost as fast as the previous rise. Yet this is not Covid-19 but the Great Plague that struck London in 1665. The main qualitative difference is seen in the small gap between the curves of plague death numbers and deaths by all causes. The Covid death toll never even accounted for half the total number of deaths in the UK, whereas in the worst week of the Plague, 19–26 September 1665, London registered 8,297 burials, of which 7,165 (86%) were from the plague.

Daniel Defoe's *Journal of the Plague Year*, though written more than 50 years after the period it depicts, gives an accurate account of the devastation caused in London, which has great resonance with the events of 2020.

In 2020, one of the factors that made it so difficult to cope with Covid-19 was the presence of symptomless carriers who could spread the disease without even knowing they had it themselves. Defoe had written this about the 1665 plague:

the danger was spreading insensibly, for the sick could infect none but those that came within reach of the sick person; but that one man who may have really received the infection and knows it not, but goes abroad and about as a sound person, may give the plague to a thousand people, and they to greater numbers in proportion, and neither the person giving the infection or the persons receiving it know anything of it, and perhaps not feel the effects of it for several days after.

In 2020, great attention was given to coming out of lockdown. The worry was that, with the end of social isolation, infection rates could rise again making matters as bad as before if not worse. Defoe also wrote of this:

Even as soon as the first great decrease in the bills [of mortality] appeared, we found that the two next bills did not decrease in proportion; the reason I take to be the people's running so rashly into danger, giving up all their former cautions and care, and all the shyness which they used to practise, depending that the sickness would not reach them – or that if it did, they should not die ... The physicians opposed this thoughtless humour of the people with all their might ... advising the people to continue reserved, and to use still the utmost caution in their ordinary conduct, notwithstanding the decrease of the distemper, terrifying them with the danger of bringing a relapse upon the whole city, and telling them how such a relapse might be more fatal and dangerous than the whole visitation that had been already; with many arguments and reasons to explain and prove that part to them, and which are too long to repeat here. But it was all to no purpose; the audacious creatures were so possessed with the first joy and so surprised with the satisfaction of seeing a vast decrease in the weekly bills, that they were impenetrable by any new terrors, and would not be persuaded but that the bitterness of death was past.

Just as a lockdown was ordered in 2020, the 1665 plague also resulted in severe restrictions ordered by the Lord Mayor of London. Among many others, these included the following:

That all plays, bear baitings, games, singing of ballads, buckler play, or such like causes of assemblies of people, be utterly prohibited, and the parties offending severely punished by every alderman in his ward.

That all public feasting, and particularly by the companies of this city, and dinners in taverns, alehouses, and other places of public entertainment, be forborne till further order and allowance, and that the money thereby spared be preserved, and employed for the benefit and relief of the poor visited with the infection.

Finally, before returning from 1665 to the present day, we should take note of Samuel Pepys' reference to the plague in his diary, where he mentions 'nobody but poor wretches in the streets', 'no boats upon the River', 'fires burning in the street' to cleanse the air and 'little noise heard day or night but tolling of bells' which were rung in London to indicate the burial of plague victims.

Pepys also showed a seventeenth-century version of our modern concerns about the results the pandemic can have on business, particularly the effect on wig-makers: 'It is a wonder what will be the fashion after the plague is done as to periwigs, for nobody will dare to buy any haire for fear of the infection – that it had been cut off of the heads of people dead of the plague.'

Fast forward to the end of 2019:

On the last day of that year, China informed the World Health Organization (WHO) about a mysterious pneumonia that had been identified in 41 people in the city of Wuhan. Most had some connection with the Huanan Seafood Market, where it was suspected that the infection had started. The following day, 1 January 2020, that market was closed down.

Events began to move rapidly. On 7 January, Chinese authorities identified a new type of coronavirus, which they called 'novel coronavirus' or nCoV, and on 11 January, the first death was recorded. The first case outside China was reported in Thailand on 13 January and other countries followed quickly: France reported Europe's first case on 17 January, the USA followed on 20 January, and by the end of March the virus had spread to at least 170 countries worldwide, causing over 50,000 deaths.

Attempts to stop, or at least decrease, the pace of spread of the infection were rapidly put in place in China. On 23 January, Wuhan was placed in quarantine, with nobody permitted to leave or enter the city without permission.

On 30 January, the WHO declared a global public health emergency and on 11 February it announced that the disease caused by the new coronavirus would be called Covid-19 (the figure indicating the year of its appearance).

On 21 February, an outbreak began in Italy, rapidly becoming Europe's worst, and on 8 March that country imposed a national quarantine, restricting all its 60 million people to their own homes except for essential purposes. Three days later, on 11 March, the WHO declared the Covid-19 outbreak to be a pandemic. By the end of March, more than a third of the world's population were under some form of lockdown.

Only time would tell whether this was a brutal necessity or a massive over-reaction and that is where the maths comes in. At times such as this, policy-makers must rely on mathematical models.

Mathematical Models

A mathematical model is a description of a system, in mathematical language, to study the effects of different components of that system and to make predictions about what will happen in the future.

Estimating the cost of a large project, or deciding how many hospital beds a city requires, or how much tax people should pay, all rely on mathematical models. In this computer age, these models have become more and more sophisticated, taking into account increasing numbers of variables. In some cases, the effect of each individual variable can be accurately assessed. But, in other cases, the model has to be based on incomplete information, and all that can then be done is to wait to see what happens and adapt the model accordingly. The coronavirus pandemic of 2020 was definitely one of those other cases.

As spokespersons for the British government never tired of saying: 'We are being guided by the science.' The trouble was that the science was necessarily guided by statistics, and the statistics were heavily dependent on accurate data, and the data gathering was incomplete and unreliable.

Building a mathematical model of the effects of coronavirus posed numerous questions to which we had little accurate idea of the answers. How many people were infected already? How quickly would the virus spread if left unchecked? What proportion of infected people would it kill? How long was the period between infection and symptoms appearing? How long between symptoms and death? Could a person who had recovered from Covid-19 catch it again, and if not, how long did the immunity last?

Those were the easy ones. To be able to make difficult decisions, we also wanted to know whether the virus discriminated between groups on the basis of age, sex, race or anything else, and did some groups suffer more severe symptoms than others? Or was it, in all such cases, an equal opportunities virus? Then there were the questions of whether our hospitals could cope and what effect large numbers of Covid-19 cases would have on the ability of staff to treat other diseases. And, perhaps most difficult of all, what effect would lockdown measures have on people's jobs and the economy?

Building all of these factors, and many more, into a mathematical model would be a massive task, even if we had all the statistics necessary to tell us the likely effects of each, which was far from the case.

Even China's experience provided limited information. By the end of April, the official figures from China confirmed 4,640 deaths, mainly in the province of Hubei of which Wuhan is the capital. The population of China is 1.39 billion, so we can calculate that the death toll from Covid-19 in China amounted to 0.0003 of 1% of the population. Applying the same percentage to the UK population would predict only 222 deaths. The great unknown, however, is not so much whether the Chinese figure was correct but what that figure would have been if the city of Wuhan had not been quarantined.

Another vital number to know is the reproductive rate, which is the average number of people an infected person will pass the disease on to. If the reproductive rate is greater than 1, the total number of infected people is bound to grow, but if it is less than 1 then the total number infected will fall, so if the pandemic is to be controlled, it is vital that the reproductive rate is reduced to below 1.

That concept captured the attention of the UK government and media and became a cornerstone of UK policy, but its appearance a few weeks into the pandemic was marred by a terminological error that hardly inspired confidence that the government truly grasped the nature of the problem it was trying to tackle.

The science of epidemiology in fact talks of two different reproductive rates. R_0 (pronounced 'R-nought') is the potential transmission rate of an infection: the rate at which a new disease can spread through a population from the start, before anyone has developed immunity or been vaccinated or gone into isolation. R_t on the other hand (alternatively known as R_e, the effective rate of reproduction) is the rate at any given time t. For at least a couple of weeks, both government sources and the media kept repeating a mantra about reducing R_0 to a value below 1, apparently unaware that it was R_t they were talking about. Eventually someone presumably told them they had got it wrong, because they all suddenly started talking simply about 'the R number', and we never heard about R-nought again.

The figures from China suggested a value around 3 for R (by which they probably meant R_0, though that was never made clear) but, to get the pandemic under control, an R value below 1 was needed. Having every sufferer passing the disease on to three others is a recipe for true exponential growth, with the number of cases tripling every time in however long it takes to pass the infection on. If $R = 1$, the number of infected people stays the same, as each sufferer either dies or gets better and is replaced by the person they infected. If R is less than 1, the number of sufferers is bound to decline.

Obtaining an accurate estimate of R, however, is not easy. First, there is the little problem of 'superspreaders'. We do not know how or why, but superspreaders can pass the virus on to as many as 100 others. Remember, the reproduction rate R is only an average, so one superspreader, infecting 100 others, and 49 carriers who do not pass it on to anyone else, still produce an average of $R = 2$.

One reason New York had such a high rate of Covid-19 has been attributed to one identified superspreader who passed the disease on to more than 100 others.

An even greater problem in calculating R is that, in most countries, we do not know how many people are infected. Quite apart from the fact that it is known that people can pass on the virus before they display any symptoms, some carriers never display any symptoms at all.

Charting hospital admission figures and fatalities may give a picture of the growth or fall of the disease, but we need the infection rates for the whole country to calculate R accurately and, even when mass testing takes place, we find ourselves relying on the accuracy of the tests, which can be doubtful at the best of times.

Even the hospital death figures pose a problem. In the UK, the published rate for Covid-19 included people who had tested positive and then died, but excluded those who displayed symptoms but were not tested, and it included those for whom factors other than Covid-19 may have been the primary cause of death. Other countries had their own policies in recording numbers of Covid-19 deaths, which made comparisons between nations difficult. In trying to establish some clarity in this respect, the WHO changed its definition of a Covid-19 death at least a dozen times, which may well have confused the matter further.

Finally, we should mention the concept of 'herd immunity', which ought to be built into any mathematical model. The idea is that if enough of the population are immune to a disease, its spread will be stopped. If two out of every three people were immune to Covid, and that R_0 figure of 3 is correct, then of the three people that an average carrier might infect, two will be immune, so only one will contract the disease, effectively reducing R_0 to 1.

The trouble is that there are basically only two ways to gain immunity: either by vaccination or, in some cases, by having had the disease. Vaccines, however, tend to take at least a year to develop and test, while the development of herd immunity in people who have recovered from the disease also has serious problems: it could easily take even longer; it might not last long; and in any case there were great doubts about whether people who had already suffered Covid-19 could catch it again or spread it.

In the absence of answers to so many vital questions, it is not surprising that the countries that achieved the best results in reducing Covid-19 rates were the ones that imposed the most stringent lockdowns the fastest. Only when the lockdowns are cautiously dismantled measure by measure would it be possible to begin to determine the effects of each measure on infection rates and know the right values to put in our mathematical models.

In May 2020, I heard two reports on BBC news announcing that the calculated value of R was 0.71 and 0.75. Given the difficulties in assessing the figures needed to calculate R, this looked suspiciously like a highly spurious degree of accuracy.

Until April 2020, testing for Covid-19 in the UK was limited mainly to people in hospital with clear symptoms of the disease and to some hospital staff. This was a highly restricted sample, giving almost no information about the level of infection in the population as a whole. In the last week of April, the Office for National Statistics (ONS) began to study a random sample of people to be able to make a valid assessment of the spread of the disease. Its first such study, of tests taken on 26 April, estimated that 0.41% of the population were infected, but applying a 95% confidence level to their results, they admitted that the true figure could be anywhere between 0.19% and 0.74%.

By 7 June, their published figure for infections had gone down to 0.06% (or, with 95% confidence, between 0.03% and 0.11%) of the population. This meant that, with 95% confidence, we could say there were about 40,000 carriers of the disease in the UK, but it may have been anywhere between 20,000 and 73,000. At the same time, the reproduction rate was asserted to be between 0.6 and 0.8, yet how such a narrow range could be justified when we were so vague about the number of infected people was not explained.

From the last week in June, the UK began relaxing its lockdown measures though this seemed to be motivated at least as much by economic considerations as the Covid-19 health data. Crucially, concerns had been growing that the costs of lockdown, both financially and to the nation's mental health, could be causing as much damage as the pandemic itself. The figures, however, demonstrated that the UK had reached its peak of infections just before lockdown had been imposed and the death rates had been declining ever since, so a gradual relaxation seemed a gamble worth taking, with the option of re-imposing it if things went badly wrong.

By the end of June, the number of Covid-19 deaths in the UK had reached 43,000 and the total number of deaths over the period of the pandemic was estimated to be 65,000 above the expected seasonal number. The 22,000 extra deaths were attributed to the huge demands Covid-19 placed on the National Health Service, which severely limited its capacity to deliver other treatments.

By the end of July 2020, the Covid-19 death rate in the UK had reached 695 per million of population. Of all the countries in the world, only Belgium had a higher per capita death rate. Even the United States (473 per million) and Brazil (449 per million) had significantly lower rates, though those countries were widely portrayed as basket cases in their attitude to Covid. With the hindsight of widespread collection and detailed analysis of data, the deficiencies of Britain's approach became clear.

Germany (where the per million death rate was 110) had instituted lockdown when the total number of deaths from Covid-19 was 86; the UK delayed it until 359 had died. Between 1 January and the time of the UK lockdown, 16 million people had entered the country, with 20,000 every day arriving from Spain alone. Genetic analysis of the virus showed that at least 1,365 people had brought it into the country, which was why it became so widespread so quickly. Furthermore, we had little idea what was happening, since testing and attempts at contact tracing were rather bizarrely stopped in the middle of March.

If lockdown had been imposed in the UK even a week earlier, it has been estimated that it would have saved the lives of between half and three-quarters of those who died.

The regulations concerning lockdown were also inconsistent. The World Health Organisation recommended that people should stay one metre apart, but the UK insisted that two metres was the correct figure. Frankly, we knew too little about how the disease was spread to say which was right. Nobody seemed quite sure whether face masks were a good idea either. And some countries closed their schools, while others kept them open.

As David Spiegelhalter pointed out, the disease was 'unbelievably safe for children' and even among anyone under-35, they were more likely to die in a road accident than from Covid-19. The worry, however, was that children could contract the disease and spread it to other more vulnerable groups. Even by the time lockdown was relaxed, however, there was little information on the extent to which children could spread the disease because the UK's track-and-trace system was still in its infancy.

What the figures did show, however, was that the chance of dying from Covid-19 once the disease has been contracted increases dramatically with age. Even a person in their 60s suffering from Covid-19 had only a 2% chance of dying from it, though this chance rose to around 10% for someone aged 90. Remarkably, at all ages, contracting Covid-19 doubled your chance of dying in the next year.

We have slowly discovered more about the spread of this disease, but governments have had to make decisions based on whatever limited information is available to them. Those who, with hindsight, got it right can shout, as in Chapter 5, 'Saved You!', while those who are seen to have misjudged things badly play a different game. This one is called 'It wasn't our fault'.

Clint Eastwood in the film *Magnum Force* said, 'A man's got to know his limitations.' As the Covid-19 saga confirms, a mathematical modeller has to know the limitations of his numbers too.

In this case, we shall only know with hindsight what approach would have been best to deal with the pandemic. Many of the lessons of the Great Plague of 1665 had clearly not been learnt 350 years later. Let us hope that hindsight does not take so long this time.

Bibliography

Some added notes on sources referred to in this book. Items are listed in the order they appear.

Introduction

Levermann, N. et al., 'Feeding Behaviour of Free-Ranging Walruses with Notes on Apparent Dextrality of Flipper Use', *BMC Ecology* (2003)

Kaplan, J.D. et al., 'Behavioural Laterality in Foraging Bottlenose Dolphins (*Tursiops truncatus*)', *Royal Society Open Science* (2019)

Chapter One: The Number of Our Days

Montagu, J.D., 'Length of life in the ancient world: a controlled study', *J. Royal Society of Medicine* (1994)

Piccioli, A., Gazzaniga, V., & Catalano, P., 'Bones: Orthopaedic Pathologies in Roman Imperial Age', Springer (2017)

Chapter 2: Surveying the Scene

Galton, F., 'Regression Towards Mediocrity in Hereditary Stature', *Journal of the Anthropological Institute* (1886)

Chapter 3: Risk and Behaviour

Kahneman, D., *Thinking, Fast and Slow*, Farrah, Straus and Giroux (2011)

Tversky, A. and Kahneman, D., 'The Framing of Decisions and the Psychology of Choice', *Science* (2010)

Thaler, R.H. and Johnson, E.J., 'Gambling With the House Money and Trying to Break: The Effects of Prior Outcome on Risky Choice', *Management Science* (1990)

Birnbaum, M., 'New Paradoxes of Risky Decision Making', *Psychological Review* (2008)

Tversky, A., 'Intransitivity of Preferences', *Psychological Review* (1969)

Blalock, G., Kadiyali, V. and Simon, D.H., 'Driving Fatalities After
9/11', *Applied Economics* (2009)
Slovic, P., 'Perception of Risk', *Science* (1987)

Chapter 4: Mathematics of Sport
Bar-Eli, M. and Azar, O.H., 'Penalty Kicks in Soccer: An Empirical
Analysis of Shooting Strategies and Goalkeepers' Preferences',
Soccer and Society (2009)
Christenfeld, N., 'What Makes a Good Sport', *Nature* (1996)
Magnus, J.R. and Klaassen, F., 'Testing Some Common Tennis
Hypotheses', Tilburg University, Center for Economic Research,
Discussion Paper (1996)
Klaassen, F. and Magnus J.R., 'Analysing Wimbledon: The Power of
Statistics', OUP (2014)
Borghans, L., 'Keuzeprobleem op Centre Court' (A choice problem
on centre court), *Economisch Statistische Berichten* (1995)
Walker, M. and Wooders, J., 'Minimax Play at Wimbledon', *The
American Economic Review* (2001)
Palacios-Huerto, I., 'Professionals Play Minimax', *Review of
Economic Studies* (2003)
Audas, R., Dobson, S. and Goddard, J., 'Team Performance and
Managerial Change in the English Football League', *Economic
Affairs* (2008)
Hope, C., 'When should you sack a football manager?', *Operational
Research Society* (2003)
Stigler, S. and Stigler, M., 'Skill and Luck in Tournament Golf',
Chance (2018)
Gilovich, T., Vallone, R. and Tversky, A., 'The Hot Hand in
Basketball: On the Misperception of Random Sequences',
Cognitive Psychology (1985)
Bocskocsky, A., Ezekowitz, J. and Stein, C., 'The Hot Hand: A
New Approach to an Old "Fallacy"', MIT Sloan Sports Analytics
Conference (2014)

Chapter 5: Saved You!
Smith, A., *The Wealth of Nations*, Strathan and Cadell (1776)
O'Rourke, P.J., *On the Wealth of Nations*, Grove Atlantic (2007)
Booker, C. and North, R., *Scared to Death*, Continuum (2007)

Maor, M., 'Policy Over-Reaction', *Journal of Public Policy* (2012)

Kahan, D.M. et al., 'Motivated Numeracy and Enlightened Self-Government, *Behavioural Public Policy*, Yale Law School (2013)

Adams, J., *Risk*, University College London Press (1995)

Chapter 6: Numbers Large and Small

Fowler, H., 'Modern English Usage', OUP (1926)

Tversky, A. and Kahneman, D., 'Belief in the Law of Small Numbers', *Psychological Bulletin* (1971)

Chapter 7: The Insignificance of Significance

Bakan, D., 'On Method: Toward a Reconstruction of Psychological Investigation', Jossey-Bass behavioural science series (1967)

Fisher, R.A., 'Statistical Methods for Research Workers', Oliver & Boyd (1925)

Gill, J., 'The Insignificance of Null Hypothesis Significance Testing', *Political Research Quarterly* (1999)

Lambdin, C., 'Significance Tests as Sorcery', *Theory and Psychology* (2012)

Chapter 8: Cause and Effect

Nietzsche, F., *Twilight of the Idols*, Naumann Verlag (1889)

Smith, G.D., 'Sex and Death: Are They Related?', *British Medical Journal* (1997)

Siminosky, K. and Bain, J., 'The Relationships Among Height, Penile Length and Foot Size', *Annals of Sex Research* (1993)

Wason, P.C., 'On the failure to eliminate hypotheses in a conceptual task', *The Quarterly Journal of Experimental Psychology* (1960)

Messerli, F.H., 'Chocolate Consumption, Cognitive Function and Nobel Laureates', *New England Journal of Medicine* (2012)

Skinner, B.F., '"Superstition" in the pigeon', *Journal of Experimental Psychology* (1948)

Campbell, D.E. and Beets, J.L., 'Lunacy and the Moon', *Psychological Bulletin* (1978)

Rotton, J. and Kelly, I., 'Much Ado About the Full Moon: A Meta-Analysis of Lunar-Lunacy Research', *Psychological Bulletin* (1985)

Näyhä, S., 'Lunar Cycle in Homicides: A Population-based Time Series Study in Finland', *British Medical Journal* (2018)

Ioannidis, J., 'Why Most Published Research Findings Are False', *PLOS Medicine* (2005)

Chapter 10: Chaotic Butterflies
Lorenz, E., 'Deterministic Nonperiodic Flow', *Journal of the Atmospheric Sciences* (1961)
Lorenz, E., 'Predictability: Does the Flap of a Butterfly's Wings in Brazil Set Off a Tornado in Texas?', Amer. Assoc. for the Advancement of Science, 139th Meeting (1972)

Chapter 11: Torpedoes, Toilets and True Love
Gardner M., 'Mathematical Games', *Scientific American*, February issue (1960)

Chapter 12:
Wong, W. et al., 'Redefining the Ideal Buttocks: A Population Analysis', *Journal of Plastic and Reconstructive Surgery* (2016)

Chapter 13: Monkey Maths
Locke, J., 'An Essay Concerning Human Understanding', Thomas Basset (1689)
Chen, K. and Santos, L., 'The Evolution of Our Preferences: Evidence from Capuchin Monkey Trading Behavior', *Cowles Foundation Discussion Paper* (2005)

Chapter 14: Pandemic Pandemonium
Defoe, D., *Journal of the Plague Year*, E. Nutt (1722)
Spiegelhalter, D., 'How much "normal" risk does Covid represent?', Winton Centre for Risk and Evidence

Acknowledgements

Numbers underlie almost every aspect of our lives and writing a book about the misuse and misinterpretations they so often provoke was a fascinating project even before Covid-19 gave a perfect example of everything I wanted to talk about. This book's production, of course, was not exempt from the massive disruption caused by the coronavirus pandemic and I must express my gratitude to everyone at Atlantic Books for nurturing this work through such a difficult period.

Special thanks are due to my editor, James Nightingale, for his encouragement and for making several suggestions that have improved the clarity of my writing and, I hope, made it easier to read. I am also indebted to my copyeditor, Mairi Sutherland, and proofreader, Ian Greensill, whose meticulous attention to detail revealed some arithmetical and logical errors that would have been very embarrassing. I must stress, however, that any such mistakes that may remain are all my own work.

Finally, back on the Covid-19 trail, I should like to thank all those writers, statisticians and mathematicians who made such valiant efforts to explain what was really going on, in the face of obfuscation and evasion by various world governments. In the case of the UK, I must give special praise to Sir David Spiegelhalter, Winton Professor of the Public Understanding of Risk at Cambridge, and Tim Harford, of BBC Radio 4's *More or Less* programme, both of whom have been outstanding in bringing clarity and honesty to the mire of statistics and disinformation surrounding this pandemic.

Index

(Bold page numbers
indicate where
definitions of
mathematical and
statistical terms may
be found.)

accident black spot, 40,
112–3
Adams, John, 112–4
Adams, John Quincy,
33–4
Advertising Standards
Agency, 177
Agoraphobia, 54
algebraic numbers,
226
al-Khwarizmi,
Muhammad ibn
Musa, 229
American football, 72,
233
Ancient Rome, 20,
29–32
annuity, 19–20
ants, 131, 209
Aristotle, 4, 168
Armani, 207
asbestos, 104–5
astrology, 31
Attenborough, Sir
David, 110–1
averages, **21–2**, 23, 48,
73, 86
Audas, R., 81
Azar, O.H., 71

babies, 11, 24, 25, 28,
98, 147
Bakan, D., 133
Bar-Eli, M., 71
baseball, 72
basketball, 95–6
bats, 11, 133
bathmophobia, 58
Bayes, T., 148
Bayes' Theorem, 148,
149–51
beauty, 42–3, 235
Becker, B., 76
Beets, J.L., 169
bell curve, **87**, 88–9,
90
Bezos, Jeff, 125
Bible, 13
billionaires, 125
Birnbaum, M. 61
Blackstone, Sir
William, 168
Bocskocsky, A., 96
Booker, Christopher,
103–5
Borghans, L., 76
Bradbury, Ray, 202–3
breast cancer, 15
breast implants, 162–3
Breslau, 18–20, 25–6
British Crime Survey,
119
brontophobia, 58
BSE, 54, 56, 104
butterfly effect, 195,
196–7, 198, 201, 208

Callaghan, James, 118
Cameron, David, 120–1
Campbell, D.E., 169
Carroll, Lewis, 3
Catastrophe Theory,
207, 208, 212, 213–5,
chaos theory, 6, 195,
196–7, 198–203,
206–8, 212, 241,
cheese, 173–4, 232
Chen, Keith, 243–4
chickens, 11, 104
child mortality, 28, 253
chimpanzees, 11, 136,
243
China, 1–2, 3, 111, 118,
133, 164, 260, 263–4
chocolate, 163–5
Christenfeld, N., 71–2
CJD, 54, 56–7, 104
Clark, Sally, 147–8
claustrophobia, 54
climacophobia, 58
clovers, 159
coin-tossing, 78, 152,
245
common cold, 161
complex numbers,
210–1,
Complexity Theory,
207, **208**, 209, 212
conditional probability,
139, 145–6, **150**
confirmation bias,
160–2, 172–3, 247
correlation, **155–6**, 157,
163–4, 167, 173–4

coronavirus, 118, 121,
133, 136, 248, 253–68
cosmetics, 176
Covid-19, 248, 257,
261–8
crime, 106, 116–7,
119–20, 146, 148,
170, 171
Croydon, 1, 2

Darwin, Charles, 158–9
Darwin, Erasmus, 159
Dawkins, Richard, 1,
194
Defender's Fallacy, 146,
148
Defoe, Daniel, 257–8
deipnophobia, 54
Descartes, René, 211
divorce, 174, 231
Djovokic, Novak, 74
DNA, 4, 136, 145–6,
148, 158
Dobson S., 81
dogs, 104, 120, 143,
186, 212, 213, 215
dolphins, 10–11
dread risk, **67**
drowning, 57
Dylan, Bob, 232

e, **224–5**, 228
Eastwood, Clint, 270
Einstein, Albert, 158,
233, 234
emails, 1, 131, 191
Empedocles, 4
empiricism, 4, 5
entrails, 16
equation, 92, 210, 211,
222, 226, 228, **233–5**,
EU Referendum, 116
Euler, Leonhard, 224–5

Everett, Daniel, 242
exchange rates, 3, 124,
125
exponential growth,
249–50, 251–4, 264
extrapolation, **16**
Ezekowitz, J., 96

Federer, Roger, 74
fertility, 253
Fisher, Ronald, 138–40
fluctuations, 3, 73, 95,
115, 203
Ford, Henry, 124
formula, **232–5**, 236–
40
Fowler, Henry, 123
fractal, 203
full moon, 167–71

Galton, Francis, 39–40
gambler's fallacy, 128,
129, 130
gambling, 187, 246
Game Theory, **77–8**,
130
Gardner, Martin, 218
Gazzaniga, Valentina,
30
genetics, 4, 5, 158
Gill, Jeff, 142, 144–5
Gilovich, Thomas, 95
giraffes, 11
Gleick, James, 200
global warming, 105,
108, 110
Goddard, John, 81
gold, 134–5
golf, 72–3, 83, 91–2, 94
gorillas, 11
Grand National, 235
Great Plague, 256–7,
259–60, 270

Graunt, John, 20
Gucci, 207
Gung, Wang, 217

Halley, Edmond, 18–21,
24–6
handedness, 8–11
happiness, 155–6
Harris, 'Bumper', 54
Hauer, Ezra, 113
Healey Denis, 124
heights of US
presidents, 90–1
Hermite, Charles, 228
honey bees, 174, 243
Hope, Chris, 82–3
hot hand fallacy, 95–6
Howell, Denis, 118
HS2 high speed rail,
122, 125–6

Ikea, 165
infant mortality, 23, 24,
26, 118
Internet, 36, 99, 188,
189, 208, 251, 252
intimacy deficit, 134,
136–7
Ioannidis, John, 172–3
IPCC, 108, 109
irrational number,
226–7

Jackson, Andrew, 33
jelly beans, 140–1
joke, formula for
perfect, 237–8

Kahan, Dan, 106
Kahneman, D., 53,
58–9, 62, 97, 122,
129, 244, 246
Kelly, Ivan, 169–70

Klaassen, F., 75–6

ladders, walking under, 159
Lambdin, Charles, 144–5
Law, John, 124
Law of Large Numbers, **126–7**, 128
Law of Small Numbers, 129–30
life expectancy, 1–32
lifespan, 15, 23, 26, 28–30
lifts, 54, 166, 167
lightning, 57, 58
Lincoln, Abraham, 90
Lindemann, Ferdinand von, 228
linear growth, **249**, 251, 253
lions, 11, 243
lockdown, 254, 255, 258, 259, 261, 263, 267–70
Locke, John, 241
Lorenz, Edward, 198–203
loss aversion, 246
lottery, 55–57, 99–101
luck, 72–3, 91–4, 113, 158–60, 199
lunacy, 167–70, 172
lung cancer, 104

Madison, James, 90
Maor, Moshe, 105
Magnus, Jan R., 75, 76
Mandelbrot, Benoit, 203
margin of error, **50–1**
mathematical models, **261–2**, 267

May, Sir Robert, 194
May, Theresa, 34
Mendel, Gregor, 4
Merrilees, Philip, 201
Messerli, Franz H., 164–5
meteorophobia, 70
Milky Way, 131, 132
millennium bug, 104
minimax, 79, 80
mobile phones, 134–5, 137
monkeys, 241–7
Monroe, Marilyn, 42
Morgan, Edward, 26
motivational distortion, 44, **45**
Munroe, Randall, 140
murder, 17, 57, 147, 168, 171, 227

Nash, John, 78
Näyha, Simo, 170
Neumann, John von, 77, 79–80
Newton, Isaac, 4, 158, 198–200, 211
Nietzsche, Friedrich, 154
North, Richard, 103–5
Nova Scotia, 9
Null Hypothesis, 139, 142, **143**, 144–5
numbrella, 4, 6

Office for National Statistics, 21, 24, 25, 26, 47, 231, 267
opinion polls, 1, 33–52, 111
O'Rourke, P.J., 102
Ovid, 198

Palacios-Huerta, Ignacio, 80
Paris Climate Agreement, 109
parrots, 11
passive smoking, 104
Pascal, Blaise, 55
Pascal's wager, 55
Patel, Priti, 116
penalty kicks, 71, 80, 81
Pepys, Samuel, 259, 260
Petty, William, 20
percentages, errors involving, 50, 175–85, 241
pigeons, 165–7
pigs, 105
plastic straws, 111
Pliny the Elder, 168
polar bears, 11–12
Popes, 30
Popper, Karl, 158–9, 162
porridge, 178
Post Traumatic Shock Disorder, 67
poverty, 2, 121, 156
projection, 16, 25, 27
Prosecutor's Fallacy, 146
Prospect Theory, 62
proton, 132
Pythagoras, 210, 227

quantum mechanics, 244
Quantum Theory, 195, 211
quark, 132

rainfall, 85, 87, 88, 89, 118,

rational numbers, 226–7
Refuge charity, 135
regression to the mean, 38, 39–41, 113
reproductive rate, 263–4
Révész, Pál, 99
risk, 31, 46, 48, 53–70, 104, 112–5, 147, 245–7
road deaths, 114
Rockefeller, John D., 124
Rotton, James, 169–70
Rutherford, Ernest, 195
Rylance, Mark, 193

salmonella, 104
sample size, 9, 140, 169, 172, 176–7
Sampras, Pete, 76
Santos, Laurie, 243–4
Sautoy, Marcus du, 194
Schrödinger, Erwin, 195
seat belts, 115
sex, 46, 48–9, 154–5, 157, 179, 180, 188, Shakespeare, William, 213
shoes, 156
SIDS, 147–8
significance, 133–153, 138
skewness, 88–9
Skinner, B.F., 166
Slovic, Paul, 69, 70
Smith, Adam, 102

Snipes, Wesley, 197, 200
soccer, 71, 80, 81, 94
Spiegelhalter, David, 248, 269
standard deviation, 83, 84–5, 86, 89–94
Stein, Carolyn, 96
Stigler, Margaret, 91–3
Stigler, Stephen, 91–3
stratified sampling, 35–7, 36
superstition, 4, 5, 33, 100, 158, 159, 161, 165–6, 171–2

tarot cards, 16
Tate Modern, 207
tennis, 74–6, 78–80, 97, 130
terrorism, 66, 67, 68
Thaler, Richard, 60
Thunberg, Greta, 110
toads, 11
toilets, 217–9
toothpaste, 177–8
traffic accidents, 66, 70, 101, 112–5, 176
train punctuality, 184–5
transcendental numbers, 226, 228
transitivity, 62–5
Trump, Donald, 44, 58, 110, 191
Turing, Alan, 229
Tversky, Amos, 58–9, 62–5, 95–6, 97, 129, 244, 246,

tweets, 131

Ulpian, 17–18

Vallone, Robert, 95
Venn diagram, 150
Vidal, Gore, 33
Vigen, Tyler, 173

waitresses, 153
Walker, Mark, 78–9
Walpole, Robert, 175
walruses, 6–10, 12
Wason, Peter, 160–1
werewolves, 168, 170, 171
wig-makers, 260
Wildavsky, Aaron, 70
Wilson, Giles, 239
Wooders, John, 78–9
World Economic Forum, 110
World Health Organisation, 118, 260, 269
world population, 14, 131, 187, 189, 250, 252–3
Wuhan, 260, 263

xanthophobia, 54

yottametre, 132

zettametre, 132

A Note About
the Author

William Hartston graduated in mathematics at Cambridge but never completed his PhD in number theory because he spent too much time playing chess. This did, however, lead to his winning the British Chess Championship in 1973 and 1975 and writing a number of chess books and newspaper chess columns.

When William and mathematics amicably separated, he worked for several years as an industrial psychologist specializing in the construction and interpretation of personality tests. After ten years writing a wide variety of columns for the *Independent*, he moved to the *Daily Express*, where he has been writing the Beachcomber column of surreal humour since 1998. In addition to writing about chess, he has written books on useless information, numbers, dates and bizarre academic research, including sexology.

Recently, his skills at sitting on a sofa watching television have been appreciated by viewers of the TV programme *Gogglebox*, but he has still not decided what he wants to be when he grows up.